First Edition

Applied Finance and General Statistical Analysis
with SAS Examples

David J. Moore, Ph.D.

Founder and President, Efficient Minds Consulting

http://www.efficientminds.com

2

Preface

The content of this book was initially assembled during my Ph.D. studies at the University of Tennessee as a means to keep track of lessons learned applying statistical analysis to finance and economic problems. Quite often I encountered scenarios where available documentation was limited, nonexistent, or was inadequately explained. It is my hope that this book fills in gaps in understanding for you as it has for me.

The first eight chapters of this book provide some of the fundamental statistical analyses (i.e., converting low-frequency data to high-frequency data and performing univariate statistics). From there more in-depth statistical analysis is covered in chapter nine which includes a discussion of asset pricing using GMM with detailed examples and chapter 10 with a discussion on the GRS test statistic. Chapter 11 goes even deeper with detailed discussions on stationarity; autocorrelation detection and correction; and heteroscedasticity detection, correction, and modeling, also including detailed examples. The book concludes with chapters 12, 13, and 14 that cover issues related with WRDS, CRSP, Compustat, and interest rate database utilization.

Contents

To Robert and Francelle for their lifelong support.

Chapter 1

Dataset operations

1.1 Arrays

Generating a one-dimensional array:

```
data test (drop=i);
   array a(10) A01-A10;
      * creates 10 variables A01, A02, ... A10;
   do i=1 to 10;
      a(i)=i;
   end;
run;
proc print noobs data=test;
run;
```

The following will accomplish the same thing (except for the leading zero):

```
data test (drop=i);
   array a(10);
      * creates 10 variables A1, A2, ... A10;
   do i=1 to 10;
      a(i)=i;
   end;
run;
proc print noobs data=test;
run;
```

1.2 Insignificant values

The COMPUSAT database has many different codes for missing data. Two of those codes, ".M" and ".I" are not exactly missing. Rather, they represent "not meaningful" and "insignificant". The following code replaces these codes with a more usable value of zero.

```
* --- convert not meaningful and insignificant values to zero;
data mydata;
set mydata;
* assign varaibles dataxyz to array notmiss;
array notmiss (i) data1 data4 data5 data6 data9 data12 data13
data14 data15 data16 data18 data19 data21 data24 data25 data30
data34 data45 data46 data60 data115 data128 data129 data149
data172 data234 data280;
do i = 1 to 27;
  if notmiss = .M then notmiss = 0;
  if notmiss = .I then notmiss = 0;
end;
```

1.3   Information

1.3.1   Number of observations and variables

```
%macro obsnvars(ds);
* Define macro obsnvars with parameter ds;
%global dset nvars nobs;
%let dset=&ds;
/* Open data set passed as the macro parameter */
%let dsid = %sysfunc(open(&dset));
/* If the data set exists, then grab the number of observations and
   variables then close the data set */
%if &dsid %then %do;
    %let nobs =%sysfunc(attrn(&dsid,nobs));
    %let nvars=%sysfunc(attrn(&dsid,nvars));
    %let rc = %sysfunc(close(&dsid));
    %put - &dset has &nvars variable(s) and &nobs observation(s).;
%end;
/* Otherwise, write a message that the data set could not be opened */
%else
    %put open for data set &dset failed - %sysfunc(sysmsg());
%mend obsnvars;
```

1.3.2   Number of non-missing observations

```
%macro nonmissobs(dsn, var);
***Find the number of nonmissing observations in data set;
PROC MEANS DATA=&DSN NOPRINT;
VAR &VAR;
OUTPUT OUT=TMP N=NONMISS;
RUN;
DATA _NULL_; SET TMP;
***Assign the value of NONMISS to the macro variable N_OBS;
CALL SYMPUT("N_OBS",NONMISS);
RUN;
%mend;
```

1.3.3 Number of missing observations

```
%macro countmiss(inds, outds, x);
proc means noprint data=&inds;
output out = &outds N(&var)=n NMISS(&var)=nmiss;
* where N() returns the number of nonmissing observations;
* and NMISS() returns the number of missing observations;
run;
data &outds;
   set &outds;
   dataset = &outds;
%mend;
```

1.4 File I/O

If using SAS from the UNIX command line then use the -NOTERMINAL option when executing SAS. PROC IMPORT and PROC EXPORT will cause UNIX SAS to freeze unless -NOTERMINAL is used.

```
%macro fileIN(filename, outds);
proc import
    datafile=&filename
    out = &outds
    replace;
run;
%mend;
%macro fileOUT(inds, fname);
proc export data=&inds
    outfile = &fname
    DBMS=csv
    replace;
run;
%mend;
```

1.5 Merging data

1.5.1 Using datastep

To merge and retain all items regardless if both input sets contribute:

```
%macro merge1 (inds1, inds2, outds, bykey);
proc sort data = &inds1;
by &bykey;
run;
proc sort data = &inds2;
by &bykey;
run;
data &outds;
    merge &inds1 (in=var1) &inds2 (in=var2);
```

```
      by &bykey;
      if var1 and var2 then match=1; else match=0;
proc sort data = &outds;
      by &bykey;
run;
%mend;
```

To merge and retain only observations in which both datasets contributed:

```
%macro merge2 (inds1, inds2, outds, bykey);
proc sort data = &inds1;
by &bykey;
proc sort data = &inds2;
by &bykey;
data &outds;
      merge &inds1 (in=var1) &inds2 (in=var2);
      by &bykey;
      if var1 and var2;
run;
proc sort data = &outds;
      by &bykey;
%mend;
```

To simply append two datasets:

```
%macro dsadd (inds1, inds2, outds);
data &outds;
      set &inds1 &inds2;
%mend;
```

1.5.2   Using proc sql

Basic merge   In this example, the dataset SAMPLE is created by matching rows in MYLIB.serank with CRSPDATA with similar GVKEY and dates.

```
proc sql;
      create table sample as select *
      from mylib.serank as l, crspdata as r
      where l.gvkey=r.gvkey and l.acyyyymm = r.yyyymm;
quit;
```

Merge and append   In this example, the dataset SAMPLE is created by matching rows in MYLIB.SERANK with CRSPDATA with similar GVKEY and dates. However, all unmatched rows in CRSPDATA are appended to SAMPLE via the RIGHT JOIN option.

```
proc sql;
      create table sample(drop=acyyyymm) as select *
      from mylib.serank as l right join crspdata as r
      on l.gvkey=r.gvkey and l.acyyyymm = r.yyyymm;
quit;
```

1.5.3   Using proc append

```
proc append base = <ds1> data = <ds2>;
run;
```

Where <DS1> represents the data set you wish to add to and <DS2> represents the dataset you wish to append.

1.5.4   Combining grand total (or mean) with original data set

```
proc means noprint data=apt;
   var mpk;
   output out = meanout (&mdrop) mean=mean;
run;
data apt;
   if _n_=1 then set meanout;
   set apt;
   mpk2 = mpk - mean;
run;
```

Note: although similar code can be found in Delwiche and Slaughter [2003], I have seen some inconsistent results. When merging summary statistics with a data set, it is helpful to add DUMMY=1 to both datasets and use DUMMY as the BYKEY. You can use the MERGE1 and MERGE2 macros of section 1.5.1 to accomplish this task.

1.6   Character data

To extract a sub-string from character data, use the SUBSTR(ARG, A, B) where ARG represents the variable name, A represents the starting character position, and B represents the number of characters to extract. For example, to take CRSP's 8-digit cusip and extract the 6-digit CNUM that is found in Compustat:

```
%let vars = date ret;
%let t0 = 1980;
%let t1 = 2000;
* retreive deleted firms from crsp events file;
data crspevents;
   set crsp.mse (keep = &vars);
   where 200 <= dlstcd <= 299 and &t0 <= year(date) <= &t1;
   cnum = substr(cusip, 1, 6);
run;
```

1.7   If-then statements

```
data ourdata;
   set ourdata;
   if answer=9 then do;
```

```
      answer=.;
      put 'INVALID ANSWER FOR ' id=;
   end;
   else do;
      answer=answer10;
      valid+1; * increment valid counter;
   end;
```

1.8   Frequency (or interval) conversion

Note: using PROC EXPAND preserves only the converted variables and the id variable unless using method=none.

1.8.1   High frequency to low frequency

Level data

```
proc expand data=monthly out=quarterly from=month to=qtr;
   convert myvar / observed=average;
   id date;
run;
```

Return data

```
proc sort dat=monret;
   by momr date;
run;
proc expand data=monret out=qtrret from=month to=qtr;
   id date;
   by momr;
   convert vwret / tin = (+1 log)
                   tout=(exp -1)
                   method=aggregate
                   observed=total;
run;
```

1.8.2   Low frequency to high frequency

```
proc expand data=datain out=dataout from=year to=qtr method=join;
   convert mpk / observed=end;
   id date;
run;
```

1.9   Transpose

Begin with dataset as follows:

```
Parameter    Estimate      StdErr     tValue      Probt
  lny0       11.79701      0.0148     797.59     <.0001
  lnk0       10.89841      0.0147     742.94     <.0001
  lnc0        7.98803      0.0125     639.07     <.0001
   m          0.021503    0.000561     38.34     <.0001
```

The following code will transpose parameter and estimate:

```
proc transpose data=pest out=transposed;
   id parameter;
   var estimate;
run;
```

and produce the following output:

```
 _NAME_          lny0        lnk0         lnc0              m
Estimate      11.79701    10.89841     7.98803       0.021503
```

Note: You can also convert from daily to monthly and monthly to yearly.

1.10   Compressing data set files

1.10.1   The easy way

There are a few ways to compress data sets in SAS:

- Using the option in the DATA step to compress a data set:

  ```
  data ssd.income (compress=yes);
  ```

- To compress all data sets created within a SAS sessions:

  ```
  options compress=yes;
  ```

- To compress a user defined library (directory) of data sets:

  ```
  libname mylib "~/pipeline/dis/lib" compress=binary;
  ```

However, this is probably necessary only with memory and storage constrained systems.

1.10.2   The hard way

First compress the file

```
compress test.dat
```

Next, use the following lines:

```
DATA test1;
FILENAME test PIPE 'zcat test.dat.Z';
INFILE test;
INPUT v1-v10;
RUN;
```

Chapter 2

Data manipulations

2.1 Growth rate time series generation

To compute a time series of growth rates from a time series of level data, use the following code:

```
proc sort data=qtrdiv;
   by momr date;
run;
proc expand data=qtrdiv out=mylib.gcf(drop=time) method=none;
   by momr;
   convert cf=gcf / transform=(log dif);
run;
```

2.2 Growth rate estimation

Following Gujarati [2003], begin with

$$Y_t = Y_0 \left(1 + r\right)^t \tag{2.1}$$

where r represents the compound growth rate over one period of time (year, quarter, month, week, day, etc.). Take the log of both sides:

$$\ln\left[Y_t\right] = \ln\left[Y_0\right] + t \ln[1 + r] \tag{2.2}$$

let

$$a_0 = \ln\left[Y_0\right] \tag{2.3}$$
$$\beta = \ln[1 + r] \tag{2.4}$$

Now (2.2) becomes:

$$\ln\left[Y_t\right] = a_0 + \beta t + e_t \tag{2.5}$$

19

where β represents the *instantaneous* growth rate. To see why β is called the instantaneous growth rate, differentiate both sides of (2.5) by t:

$$\frac{d\left(\ln\left[Y_t\right]\right)}{dt} = \frac{1}{Y}\frac{dY}{dt} = \beta \tag{2.6}$$

The compound growth rate can be computed from the instantaneous growth rate β as:

$$r = \exp\left[\beta\right] - 1 \tag{2.7}$$

In SAS, PROC MODEL can be used to arrive at β:

```
* -------------------------------------;
* Macro to compute long-run growth rates;
* -------------------------------------;
%macro gcalc(x, y);
* x = dataset;
* y = varaiable;
title2 "g_&y (Newey-west std errors)";
data gcalcin;
    set &x;
    lny = log(&y);
run;
ods listing close;
ods output parameterestimates=nw residsummary=fit;
proc model data=gcalcin;
    parms a0 beta;
    lny = a0 + beta*t;
    fit lny / gmm maxiter=500 kernel=bart;
run;
ods listing;
proc print data=nw(firstobs=2);
    id parameter;
    format estimate stderr 7.4;
run;
data fit;
    set fit;
    keep esttype equation rsquare adjrsq;
run;
proc print data=fit;
    id equation;
run;
%mend;
```

2.3   Discount rate estimation

In discrete time, the present value of future values N_t is expressed as:

$$N = \sum_{t=0}^{T}\frac{N_t}{(1+b)^t} = \sum_{t=0}^{T}\beta^t N_t \tag{2.8}$$

where β represents the discrete discount *factor* and b is the discrete periodic discount *rate*. In continuous time:

$$N = \int_0^T N[t]e^{-\theta t}dt \tag{2.9}$$

where $e^{-\theta t}$ is the continuous (instantaneous) discount *factor* and θ is the instantaneous discount *rate*. The discrete discount factor β and instantaneous discount rate θ are related by the following equations:

$$e^{-\theta} = \beta = \frac{1}{1+b}$$

$$\theta = -\ln[\beta] = -\ln\left[\frac{1}{1+b}\right]$$

2.4 Cumulative return

If converting between standard SAS dates (e.g., monthly to quarterly or quarterly to annual), the easy way can be found in Section 1.8.1.

If converting daily to weekly, or some other combination not covered in Section 1.8.1, you will have to take a few more steps. Let R_t represent the total return return at time t. The cumualtive return can be computed as:

$$CUMRET_T = \prod_{t=1}^T (1 + R_t) - 1$$

taking the natural log of both sides:

$$\ln[CUMRET_T + 1] = \sum_{t=1}^T \ln[1 + R_t]$$

which is simply the arithmetic mean of the natural log. As such, the compound growth rate can be obtained as:

$$CUMRET_T = \exp\left[\sum_{t=1}^T \ln[1 + R_t]\right] - 1$$

The associated SAS code is:

```
* ----------------;
* Cumulative return;
* ----------------;
%macro cumret(inds, outds, var, byvars);
proc sort data=&inds;
    by &byvars;
run;
```

```
* convert toal returns to log returns;
proc expand data=&inds out=ln&inds(drop=time) method=none;
   convert &var=ln&var / transform=(+1 log);
run;
* compute sum of log returns;
proc means noprint data=ln&inds;
   by &byvars; *e.g., permno date (where date is quarterly);
   var ln&var;
   output out = tmpcumret(&mdrop) sum=;
run;
* convert sum of log returns to total return;
proc expand data=tmpcumret out=&outds(drop=time) method=none;
   convert ln&var=&var / transform=(exp -1);
run;
%mend;
```

2.5   Moving averages

```
proc expand data=datain out=dataout method=none;
   id date;
   convert mpk / transform=(movave 4);
   convert gcf / transform=(movave 4);
run;
```

2.6   Inflation adjustment

2.6.1   Level data

2 step process:

1. Get PCE deflators from www.bea.gov

2. Run the following code which uses the MERGE2 macro from Section 1.5.1:

```
* --------------------------;
* Deflate quarterly level data;
* --------------------------;
%macro deflqlvl(inds, outds, var);
%put Deflate quarterly level data...;
%put - merge deflation data with &inds...;
%merge2(mylib.pceqtr, &inds, merge2out, date);
%put - compute deflated time series; data &outds;
   set merge2out;
   &var = (99.296/defl)*&var;
   * note 99.296 is the 2000Q1 value;
   drop defl;
run;
%put done.;
%mend;
```

2.6.2 Return data

```
%merge2(pcedata, inds, outds, date);
proc sort data=outds;
   by date;
run;
data outds;
   set outds;
   ldefl = lag(defl);
   realret = (ldefl / defl)*(1+ret)-1;
run;
```

Chapter 3

Date manipulations

3.1 Date Formats

format	example
YYMMN6.	198304
YYMMDD8.	19830401
DATE9.	01APR1983
YYQ6.	1991Q1
YYQC6.	1991:1

3.2 sas to integer

3.2.1 sas to yyyyq

```
data _null_;
   yyyyq = trim(year(date))||"Q"||trim(qtr(date));
run;
```

3.3 Integer to sas

3.3.1 yyyy to SAS

```
%macro yyyy2sas(x);
data &x;
   set &x;
   format date date9.;
   chardat = put(100*yyyy+12, 6.);
   datey = trim(chardat);
   date = input(datey, yymmn6.);
   drop yyyy datey chardat;
```

```
    run;
    %mend;
```

3.3.2   yyyyq to sas

```
%macro yyyyq2sas(x);
data &x;
    set &x;
    format date yyq6.;
    * convert interger to character;
    chardat = put(yyyyq, 5.); /* convert int to char */
    yyyy = substr(chardat,1,4); /* year */
    q = substr(chardat,5,1); /* quater */
    dateq = trim(yyyy)||"Q"||trim(q);
    date = input(dateq, yyq6.);
    drop yyyyq yyyy q dateq chardat;
run;
%mend;
```

3.3.3   yyyymm to sas

```
data _null_;
    intdat = 198304;
    chardat = put(intdat, 6.); /* convert int to char */
    idatum = input(chardat, yymmn6.); /* convert to sas date */
    put chardat = ; /*original char value */
    put datum = ; /* internal representation */
    put datum = yymmn6.; /* 6 digit representation */
    put datum = date9. ; /* normal date representation */
run;
```

which leads to the following output:

```
chardat=198304
datum=8491
datum=198304
datum=01APR1983
```

3.3.4   yyyymmdd to SAS date

```
data x (drop=chardat yyyymmdd);
    set x;
    chardat = put(yyyymmdd,8.);
    date = input(chardat, yymmdd8.);
    format date yymmdd8.;
run;
```

3.4 sas to sas / integer to integer

3.4.1 yyyyq to yyyymm

```
data mpkpc;
   set mpkpc;
   format date yymmn6.;
   chardat = put(cyyyyq, 5.);
   charyyyy = substr(chardat,1,4); /* year */
   q = substr(chardat,5,1); /* quater */
   if q = 1 then mm = "03";
       else if q = 2 then mm = "06";
       else if q = 3 then mm = "09";
       else if q = 4 then mm = "12";
   date = input(trim(charyyyy)||trim(mm), yymmn6.);
   keep date mpk;
run;
```

3.4.2 sasmonthly to sasquarterly

```
data new;
   set orig;
   format date yyq6.;
   date = yyq(year(date),qtr(date));
run;
data new;
   set orig;
   by date;
   if first.date;
run;
```

3.4.3 sasquarterly to sasmonthly

Next edition? :)

Chapter 4

Macros

4.1 Global macro variables

To define a global macro variable you can use:

```
%let globvar = 5;
```

Or,within a data step:

```
call symputx('globvar', dsvarname, 'g');
```

4.2 If-then statements

Chapter 5

Output Delivery System (ODS)

5.1 ODS example: parameter estimates

See the following annotated code:

```
* turnoff listing;
ods listing close;
* define what/where to output
ods output parameterestimates=nw;
* execute process;
proc model...
```

5.2 Latex tables

Here is an example:

```
ods select parameterestimates residsummary;
ods tagsets.tablesonlylatex file="tex/cd.tex" (notop nobot) newfile=table;
proc model data=mylib.qtrlog;
    parms a0 a1;
    instruments k ;
    y = a0 + a1*k ;
    fit y / gmm maxiter=500 kernel=(bart,4,0);
    run;
ods tagsets.tablesonlylatex close;
```

5.3 R^2 calculations

The following table lists various measures from regression output with names used in formulas to follow.

Description	SAS name	DJM name
Residual (error) sum of squares	SSE	RSS
Model (explained) sum of squares	?	ESS
Total sum of squares	?	TSS
total R^2	R^2_{tot}	R^2
regression R^2	R^2_{reg}	?
adjusted R^2	R^2_{adj}	$\overline{R}^2$

The total sum of squares may not be explicitly present in regression output (e.g., proc autoreg) but it can be calculated from the residual sum of squares measure.

$$TSS = \frac{RSS}{1 - R^2_{tot}}$$

Once TSS is obtained, ESS can be computed as $ESS = TSS - RSS$. The adjusted R^2 can be computed using:

$$\overline{R}^2 = 1 - \frac{RSS/(n-k)}{TSS/(n-1)}$$

where n is the number of obserations and k is the number of parameters.

Chapter 6

proc iml

- From SAS dataset to PROC IML matrix

```
use dataset-name;
read all var {var1 var2...} into matrix-name;
```

- From IML vector to SAS dataset

```
create dataset-name from matrix-name;
append from matrix-name;
close dataset-name;
```

- From IML variables (vectors or matricies) to SAS dataset

```
create dataset-name var {var1 var2...};
append;
close dataset-name;
```

Chapter 7

proc model

7.1 Equation estimation

7.1.1 General and normal form systems of equations

General form:

$$\epsilon_{y,t} = lny_t - a_y - m_y t$$
$$\epsilon_{k,t} = lnk_t - a_k - m_k t$$
$$\epsilon_{c,t} = lnc_t - a_c - m_c t$$

Normal form:

$$lny_t = a_y + m_y t + \epsilon_{y,t}$$
$$lnk_t = a_k + m_k t + \epsilon_{k,t}$$
$$lnc_t = a_c + m_c t + \epsilon_{c,t}$$

In SAS terminology, $a_y, a_k, a_c, m_y, m_k, m_c$ are *parameters* (to be estimated) while lny_t, lnk_t, lnc_t, and t are *variables*.

7.1.2 Siimultaneaity and cross-equation error correlation

- **SUR (joint GLS):** In the example of the previous section, there are no endogenous variables on the right hand side. However, if the error terms are cross-correlated, Seemingly Unrelated Regression (SUR) can be used to improve estimation efficiency in "large" samples.

- **2SLS**: Used if endogenous variables are on the RHS. Requires instruments but no assumptions about error term distributions.

- **3SLS (2SLS + SUR)**: Simultaneous equations with cross-equation error correlation.

- **FIML**: Handles simultaneity without the use of instrumental variables. However, this method assumes errors have a multivariate normal distribution.

- **GMM**: Makes use of cross-equation error term correlation.

7.2   Heteroscedasticity

7.2.1   Known error variance relationship

In the unlikely event you know the error term variance relationship *a priori*, there are at least two estimation methods available.

1. Weighted regression. The WEIGHT T option must follow the FIT statement. The weight is appplied to the squared residuals.

   ```
   proc model data=test;
       parms b1 0.1 b2 0.9;
       y = 250*(exp(-b1*t)-exp(-b2*t));
       fit y;
       weight t;
   run;
   ```

2. GLS estimation. Next edition?

7.2.2   Unknown error variace relationship

1. GMM estimation. This estimation requires instruments, however, if you specify parameters as instruments, the partial derivatives of the equations with respect to that parameter are taken as instruments. These partial derivatives should not depend on any of the parmeters to be estimated. A list of exogenous variables is sufficient as shown in Section 9.8. Also, the omisssion of instruments causes SAS to use the explanatory variables as instruments. As such, the coefficient estimates are the same as OLS but the standard errors are corrected for heteroscedasticity and auto-correlation when using KERNEL=BART.

2. HCCME. MacKinnon and White (1985) showed the third modification of the White (1980) heteroscedastic consistent-covariance matrix estimator (HCCME) performed the best. Therefore use the HCCME=3 option in the fit statement. E.g., FIT Y / SUR HCCME=3.

3. ARDCH/GARH estimation. The first two approaches (GMM and HCCME) *correct* standard errors while ARCH / GARCH *model* the error term variance relationship. More on this in Chapter 11.

7.3 Storing / retreiving models

Model equations, parameters, and other parameters can be stored using the outmodel= statement:

```
proc model data=modelin outmodel=mylib.ccapm.fhbase;
    ... model statements (including fit)
quit;
```

To retrieve a model, including the model specification and parameter estimates, use the model= statement:

```
proc model model=mylib.ccapm.fhbase; quit;
```

Or, for multiple models:

```
proc model model=(model1 model2...); quit;
```

7.4 Convergence problems/solutions

7.4.1 What to check first

1. Explicitly identify endogenous, exogenous, and instrumental variables.

2. [More to be added in future editions of this text.]

7.4.2 Local Minimum as opposed to global minimum

From the SAS documentation [SAS Institute Inc., 2008]:

"You may have converged to a local minimum rather than a global one. This problem is difficult to detect because the procedure will appear to have succeeded. You can guard against this by running the estimation with different starting values or with a different minimization technique. The START= option can be used to automatically perform a grid search to aid in the search for a global minimum."

Also from the SAS codumentation [SAS Institute Inc., 2008]:

"For example, the following statements try 5 different starting values for C: 10, 5, 2.5, -2.5, -5. For each value of C, values for A and B are estimated. The combination of A, B, and C values producing the smallest residual mean square is then used to start the iterative process."

```
proc model data=test;
   parms a b c;
   y = a + b * x ** c;
   label a = "Intercept" b = "Coefficient of Transformed X"
         c = "Power Transformation Parameter";
   fit y start=(c=10 5 2.5 -2.5 -5) / startiter itprint;
run;
```

Unfortunately, in my dissertation scenario, neither the minimization method nor the START= option helped.

Chapter 8

Univariate Statistics

8.1   Computing weighted returns

```
proc sort data=inds out=outds;
   by rank yyyymm;
run;
proc means data = outds noprint;
   by momr yyyymm;
   var ret / weight=size ;
   output out = vwretdat mean= vwret;
run;
```

8.2   Computing mean, t-stats, and p-values

```
* -------------------------------------------;
* macro to compute mean, t-stats, and p-values;
* -------------------------------------------;
%macro meanit (inds, outds, var, classvar);
proc means noprint data=&inds ;
   class &classvar;
   var &var;
   output out = &outds(drop=_type_ _freq_) mean=mean t=t probt=p;
run;
data &outds;
   set &outds;
   mean = round(mean, 0.0001);
   t = round(t, 0.01);
   p = round(p, 0.001);
run;
proc print data=&outds;
   id &classvar;
run;
%mend;
```

8.3   Histograms

Generating histograms is fairly easy using the PROC UNIVARIATE routine:

```
proc univariate data=datain noprint;
   histogram varname;
run;
```

Chapter 9

Asset pricing using GMM

9.1 Introduction

9.1.1 Central asset pricing formula

Consider the intertemporal optimization problem:

$$\max_{\xi} u\left[c_t\right] + E_t\left[\beta u\left[c_{t+1}\right]\right] \quad \text{s.t.}$$

$$c_t = e_t - p_t\xi$$
$$c_{t+1} = e_{t+1} + x_{t+1}\xi$$

where ξ represents the amount of payoff x_{t+1} purchased and e_t represents original consumption level (if none of the asset were purchased) at time t. The first order condition for the maximization problem is:

$$p_t u'\left[c_t\right] = E_t\left[\beta u'\left[c_{t+1}\right] x_{t+1}\right] \tag{9.1}$$

Equation (9.1) represents the central asset pricing formula and may be interpreted as follows:

- $p_t = $ current price of asset

- $u'\left[c_t\right] = $ marginal utility (utility per dollar) at time t

- $p_t u'\left[c_t\right] = $ loss in utility for buying asset at time t

- $u'\left[c_{t+1}\right] x_{t+1} = $ gain in utility from asset's payoff at time $t + 1$

- $\beta u'\left[c_{t+1}\right] x_{t+1} = $ current value of utility gain

9.1.2   Stochastic discount factor, marginal rate of substitution, pricing kernel, moment condition

Equation (9.1) can be rearranged to;

$$p_t = E_t \left[\beta \frac{u'\left[c_{t+1}\right]}{u'\left[c_t\right]} x_{t+1} \right] \tag{9.2}$$

Define the stochastic discount factor, m_{t+1}, as

$$m_{t+1} = \beta \frac{u'\left[c_{t+1}\right]}{u'\left[c_t\right]} \tag{9.3}$$

which:

1. is *stochastic* because consumption at time $t+1$ is unknown at time t (and is related to the payoff x_{t+1} which is also unknown);

2. is often called the *marginal rate of substitution*: the rate at which consumption at time t is foregone for consumption at time $t+1$;

3. is sometimes referred to as the *pricing kernel*, *change of measure*, or *state-price density*; and

4. has the same functional form for all assets although the correlation of x_{t+1} and e_{t+1} (the original overall consumption level) impacts asset-specific risk correction.

As such, the central asset pricing formula can be rewritten as:

$$\boxed{p_t = E_t \left[m_{t+1} x_{t+1}\right]} \tag{9.4}$$

If we divide both sides by p_t and define $R_{t+1} = x_{t+1}/p_t$ the asset pricing equation can be rewritten:

$$\boxed{1 = E_t[m_{t+1} R_{t+1}]} \tag{9.5}$$

Rearranging this a bit we have the *moment condition*:

$$E_t \left[m_{t+1} R_{t+1} - 1\right] = 0$$

9.1.3   Risk free asset

Substituting the risk-free rate into (9.5):

$$1 = E_t \left[m_{t+1} R^f_{t+1}\right]$$

since the asset is risk free, the payoff R^f_{t+1} is known at time t (e.g., think of a zero coupon government bond or T-bill) and can be brought outside the expectations operator. Note: we are talking about the *nominal* risk free rate since the *real* rate is unknown at time t (because the "true" inflation is unknown and expected inflation is unobservable). Therefore the risk free rate can be written as:

$$R_{t+1}^f = \frac{1}{E_t[m_{t+1}]}$$ (9.6)

Removing uncertainty and substituting (9.3) into (9.6):

$$\boxed{R_{t+1}^f = \frac{1}{\beta}\frac{u'[c_t]}{u'[c_{t+1}]}}$$ (9.7)

where β represents the discrete discount rate from period $t+1$ to t. For CRRA utility, $u[c] = \frac{c^{1-\gamma}-1}{1-\gamma}$ equation (9.7) becomes:

$$R_t^f = \frac{1}{\beta}\left(\frac{c_{t+1}}{c_t}\right)^\gamma$$

With DRRA utility such that $u'[c] = \exp[\gamma/\lambda c^\lambda]$, $\gamma > 0$, $\lambda > 0$:

$$R_t^f = \frac{1}{\beta}\exp\left[\frac{\gamma}{\lambda}\left(\frac{1}{c_t^\lambda} - \frac{1}{c_{t+1}^\lambda}\right)\right]$$

where γ represents the degree of risk aversion and λ represents the rate of decrease in risk aversion. See Meyer and Meyer [2005] for details on proper selection of γ and λ.

9.2 Asset pricing model fundamentals

9.2.1 Discount factor model

Asset pricing can be summarized by two equations

$$p_t = E_t[m_{t+1}x_{t+1}]$$ (9.8)

$$m_{t+1} = \beta\frac{u'[c_{t+1}]}{u'[c_t]}$$ (9.9)

where p_t is the asset price, x_{t+1} is the asset payoff, and m_{t+1} is the stochastic discount factor. Regarding the discount factor, β represents the *subjective discount factor* (patience) and $u'[c_{t+1}]/u'[c_t]$ captures the curvature of the utility function, and this curvature reflects aversion to risk.

9.2.2   Expected return-beta model

Begin with the relationship $\text{cov}[m, x] = E[mx] - E[m]E[x]$, which can be rewritten as[1]

$$E[mx] = \text{cov}[m, x] + E[m]E[x] \tag{9.10}$$

Substituting this expression, and $E[m] = 1/R_f$, into the asset pricing equation 9.5:

$$1 = \frac{E[R_i]}{R_f} + \text{cov}[m, R_i]$$
$$E[R_i] = R_f - R_f\text{cov}[m, R_i]$$
$$E[R_i] - R_f = -R_f\text{cov}[m, R_i] \tag{9.11}$$

Keep in mind that variance of an asset's return is **not** the central concern of a consumer. Rather, it is the covariance of returns with the discount factor. Therefore, only risk associated with the discount factor m generates a premium.

Returning to equation (9.11), noting $R_f = 1/E[m]$, and multiplying and dividing by $var[m]$:

$$E[R_i] = R_f + \left(\frac{-\text{cov}[m, R_i]}{\text{var}[m]}\right)\left(\frac{\text{var}[m]}{E[m]}\right)$$

Allow me to introduce two definitions:

- *Quantity of discount rate risk:* The sensitivity of asset i to discount rates, $\beta_{i,m} = {-\text{cov}[m, R_i]}/{\text{var}[m]}$, which can be obtained from a regression of R_i on m.

- *Price of discount rate risk:* $\lambda_m = {\text{var}[m]}/{E[m]}$, which is the same for all assets,

Using these definitions we arrive at the *expected-return beta* form of the asset pricing equation:

$$\boxed{E[R_i] = R_f + \beta_{i,m}\lambda_m} \tag{9.12}$$

9.2.3   Excess returns

Equation (9.12) can be expressed in excess return form:

$$E[R_i] - R_f = \beta_{i,m}\lambda_m$$
$$E[R_{ei}] = \beta_{im}\lambda_m$$

---

[1]This is the phase where higher order correlations could be introduced. I am unsure at the moment what an expression for higher order correlations might be, but if it includes $E[mx]$ then this model can certainly be expanded.

Solving backwards to the discount factor expression:

$$E[mR_{ei}] = \text{cov}[m, R_{ei}] + E[m]E[R_{ei}]$$
$$= \text{cov}[m, R_{ei}] + \frac{1}{R_f}(E[R_i] - R_f)$$
$$= \text{cov}[m, R_{ei}] + \frac{1}{R_f}E[R_i] - 1$$

Substituting $E[R_i] = R_f - R_f \text{cov}[m, R_i]$ (from equation 9.11):

$$E[mR_{ei}] = \text{cov}[m, R_{ei}] + \frac{1}{R_f}(R_f - R_f\text{cov}[m, R_i]) - 1$$
$$= \text{cov}[m, R_{ei}] + 1 - \text{cov}[m, R_i] - 1$$
$$= \text{cov}[m, R_{ei}] - \text{cov}[m, R_i]$$

Using equations (9.10)(9.5), and (9.6); and the the property of covariance of a difference, we can expand $\text{cov}[m, R_{ei}]$ as follows:

$$\text{cov}[m, R_{ei}] = \text{cov}[m, R_i - R_f]$$
$$= \text{cov}[m, R_i] - \text{cov}[m, R_f]$$
$$= \text{cov}[m, R_i] - (E[mR_f] - E[m]E[R_f])$$
$$= \text{cov}[m, R_i] - (1 - E[m]R_f)$$
$$= \text{cov}[m, R_i] - (1 - 1)$$
$$= \text{cov}[m, R_i]$$

Therefore,

$$\boxed{E[mR_{ei}] = 0} \tag{9.13}$$

.

9.3 Linear factor models

9.3.1 Setup

Utility is unobservable and therefore difficult to model. Even with a functional form of utility, consumption data is problematic. Therefore, we can approximate the stochastic discount factor:

$$m_{t+1} = \beta \frac{u'[c_{t+1}]}{u'[c_t]} \approx a_0 + a_1 f_{1,t+1} + a_2 f_{2,t+1} + \cdots + a_k f_{k,t+1} \tag{9.14}$$

where f represent risk factors or factors that are related to marginal utility growth. In my opinion, choosing factors based on returns (e.g., returns of small stocks less returns of large stocks) is an ad-hoc approach. This specification is equivalent to a multiple-beta (I will use b instead of β) representation by the following <u>cross-sectional</u> regression:

$$E_T\left[R_i\right] = \lambda_0 + \lambda_1 b_{i1} + \lambda_2 b_{i2} + \cdots + \lambda_k b_{ik} \quad i = 1 \ldots N \tag{9.15}$$

where the independent variables b_i are obtained from the <u>time-series</u> regression:

$$R_{i,t} = b_{i0} + b_{i1} f_{1,t} + b_{i2} f_{2,t} + \cdots + b_{ik} f_{k,t} + e_{i,t} \quad t = 1 \ldots T \tag{9.16}$$

9.3.2 Factor characteristics

Risk factors f have the following characteristics:

1. Market-wide influence on stock returns (b_{ik} statistically significant for all i).

2. Must influence expected return therefore have nonzero prices $\lambda_k \neq 0$.

3. Factor must be unpredictable. For instance, inflation is partially predictable therefore not an APT factor. However, the portion of inflation which is not predictable (i.e., unexpected inflation), $\text{INFL} - E[\text{INFL}]$ is an APT factor.

4. "Good" proxy for marginal utility growth - see equation (9.14).

5. Indicate "bad states."

Characteristics 1 and 2 are articulated by Fama and French (1993):

> "... variables that are related to average returns must proxy for sensitivity to common risk risk factors in returns."

If there is a risk-free asset with return R_f then $b_{i,j} = 0 \ \forall j$ since there is no risk. Therefore $\lambda_0 = R_f$.

9.3.3 Estimation

9.3.3.1 Case 1: Test assets are gross returns, factors are gross returns

Recall the three asset pricing equations:

$$0 = E\left[(a_0 + a_1 f_{1,t+1} + a_{2,t+1} + \cdots + a_k f_{k,t+1})\, R_{i,t+1} - 1\right] \tag{9.17}$$

$$E_T\left[R_i\right] = \lambda_0 + \lambda_1 b_{i1} + \lambda_2 b_{i2} + \cdots + \lambda_k b_{ik} + u_i \quad i = 1 \ldots N \tag{9.18}$$

$$R_{i,t} = b_{i0} + b_{i1} f_{1,t} + b_{i2} f_{2,t} + \cdots + b_{ik} f_{k,t} + e_{i,t} \quad t = 1 \ldots T \tag{9.19}$$

Three pieces of information can be gathered from the asset pricing equations:

1. a_k: Does factor f_k help price all assets given other factors?

2. λ_k: What is the market price of risk associated with factor k?

3. b_{ik}: How sensitive is firm i's returns to factor k?

The model can be applied to the factors themselves. As such, for factor k, $b_{ik} = 1$ and zero otherwise which leads to the relationship $E[f_k] = \lambda_k$ and equation (9.18) can be rewritten as

$$E_T[R_{ei}] = \lambda_0 + E[f_1]b_{i1} + E[f_2]b_{i2} + \cdots + E[f_k]b_{ik} + u_i \quad i = 1 \ldots N$$

9.3.3.2 Case 2: Test assets are excess returns, factors are gross returns

Let $R_{ei} = R_i - R_j$ represent excess return (R_j need not be the risk free rate). From Section 9.2.3 we know $E[m_{t+1}R_{ei,t+1}] = 0$. Furthermore

$$E_T[R_{ei}] = E_T[R_i] - E_T[R_j] \tag{9.20}$$

which eliminates the intercept from (9.15). Therefore, the trio of asset pricing equations can be rewritten as:

$$0 = E[(a_0 + a_1 f_{1,t+1} + a_{2,t+1} + \cdots + a_k f_{k,t+1})R_{ei,t+1}] \tag{9.21}$$

$$E_T[R_{ei}] = \lambda_1 b_{i1} + \lambda_2 b_{i2} + \cdots + \lambda_k b_{ik} + u_i \quad i = 1 \ldots N \tag{9.22}$$

$$R_{ei,t} = b_{i0} + b_{i1} f_{1,t} + b_{i2} f_{2,t} + \cdots + b_{ik} f_{k,t} + e_{i,t} \quad t = 1 \ldots T \tag{9.23}$$

One advantage in using excess returns is you need not rely on a constant risk free rate. Therefore, the time-varying T-bill, for example, can be used as R_j. Also note the implication of (9.22) and (9.23) is: $b_{i0} = 0 \quad \forall i$.

The model can not be applied to the factors themselves as with the previous case since factors are *excess* returns and test assets are *gross* returns.

9.3.3.3 Case 3: Test assets are excess returns, factors are excess returns

In this case, the model can be applied to the factors themselves. As such, for factor k, $b_{ik} = 1$ and zero otherwise which leads to the relationship $E[f_k] = \lambda_k$ and (9.22) can be written:

$$E_T[R_{ei}] = E[f_1]b_{i1} + E[f_2]b_{i2} + \cdots + E[f_k]b_{ik} + u_i \quad i = 1 \ldots N \tag{9.24}$$

In this case, estimates for the market prices of risk can be estimated directly without the need of cross-sectional regression:

$$\widehat{\lambda}_k = E_T[f_k] \tag{9.25}$$

9.3.3.4 When factors are not returns

When factors are not returns, the time series regression of the form (9.19) can not be used with the factor as the dependent variable. In other words, the market price of risk for factors that are not returns can not be estimated by (9.25). In this case, follow the procedure of Section 9.3.3.1 or 9.3.3.2 above (regress returns on factor, then regress average returns on sensitivities).

9.3.4   Summary: specific tests

Recall the discount factor, expected return-b, and time-series equations:

$$m_{t+1} = a_0 + a_1 f_{1,t+1} + a_2 f_{2,t+1} + \cdots + a_k f_{k,t+1}$$
$$E_T[R_{ei}] = \lambda_0 + \lambda_1 b_{i1} + \lambda_2 b_{i2} + \cdots + \lambda_k b_{ik}$$
$$R_{ei,t} = b_{i0} + b_{i1} f_{1,t} + b_{i2} f_{2,t} + \cdots + b_{ik} f_{k,t} + e_{i,t}$$

Several tests can be performed:

1. Does factor j help price assets in the presence of other factors? Test $H_0 : \quad a_j = 0$ by running GMM on the moment condition for two or more assets:

$$E[m_{t+1} R_{i,t+1} - 1] = 0$$

   or for excess returns:

$$E[m_{t+1} R_{ei,t+1}] = 0$$

   When factors are correlated, which is quite often the case, one should test $b_j = 0$ to determine if it should be included rather than $\lambda_j = 0$.

2. Is factor j priced? Alternatively, is factor j correlated with the true discount factor? Test $H_0 : \quad \lambda_j = 0$

   (a) If factors are returns, you can simply estimate $\widehat{\lambda}_k = E_T[f_k]$

   (b) If factors are *not* returns, obtain b_{ik} from time series regression of returns onto factors then run cross sectional regression to obtain $\widehat{\lambda}_k$

3. Does the collection of factors completely price assets? Alternatively, are excess returns zero after controlling for known risk factors? Test $H_0 : \quad b_{i0} = 0 \ \forall \ i$ by running the time series regression and performing joint test of $b_{i0} = 0$.

9.4   From log utility to CAPM

Consider the case of log utility:

$$u[c_t] = \ln[c_t]$$

Let p_t^W represent the price of a claim to all future consumption:

$$p_t^W = E_t \sum_{j=1}^{\infty} \beta^j \frac{u'[c_{t+j}]}{u'[c_t]} c_{t+j} = E_t \sum_{j=1}^{\infty} \beta^j \frac{c_t}{c_{t+j}} c_{t+j} = \frac{\beta}{1-\beta} c_t \qquad (9.26)$$

It is worth pausing for a moment to describe how log utility has the property that "the income effect exactly offsets the substitution effect."

- Income effect: news of higher c_{t+1} →should make claim p_t^W more valuable

- Substitution effect: higher c_{t+1} → lower marginal utility $u'[c_{t+1}]$ in period $t+1$. I.e., *substituting* period t consumption for period $t+1$ consumption results in *lower* marginal utility in period $t+1$ when c_{t+1} is expected to increase.

- Price effect (net of income and substitution effect): none, as evidenced by the absence of c_{t+1} in the final expression for p_t^W,

From (9.26) expression it can be shown the wealth portfolio return is proportional to marginal utility growth:

$$R_{t+1}^W = \frac{p_{t+1}^W + c_{t+1}}{p_t^W} = \frac{\left(\frac{\beta}{1-\beta}\right)c_{t+1} + c_{t+1}}{\left(\frac{\beta}{1-\beta}\right)c_t} = \frac{c_{t+1}\left(\frac{1}{1+\beta}\right)}{\left(\frac{\beta}{1-\beta}\right)c_t} = \frac{1}{\beta}\frac{c_{t+1}}{c_t}$$

However, $u'[c] = 1/c$ therefore:

$$R_{t+1}^W = \frac{1}{\beta}\frac{u'[c_t]}{u'[c_{t+1}]} \to m_{t+1} = \frac{1}{R_{t+1}^W}$$

Therefore the linear approximation is the familiar CAPM discount factor:

$$m_{t+1} \approx a + bR_{t+1}^W$$

Cochrane [2005] points out that linear approximations to the nonlinear discount factor are not without problems For longer time horizons, the approximation loses accuracy. Therefore, in general, it is not a good idea to apply linear approximations to longer time intervals as the error of the approximation increases with the magnitude of the factor.

So what is left to estimate? Linearizing any specification of the consumption based discount factor is questionable so how do we conduct an asset pricing tests? Since GMM can handle nonlinear regression (i.e., equations nonlinear in the parameters and/or factors), it can be applied *directly* to the consumption CAPM discount factor pricing equation. However, we still face the problem of low volatility nondurable consumption and measurement error involved in collecting consumption data, and precise utility function specification.

In sum, linear factor models, due to the linear approximation of nonlinear discount factors, are best suited for short-horizon estimations. Nonlinear factor models are better suited for longer horizon models.

9.5 GMM estimation mechanics

This section provides a brief description of the GMM estimation technique of Hansen [1982]. Method of moments (MM) and generalized method of moments (GMM) estimation is based on a set of population *moment conditions* that include data and unknown parameters. Estimates based on sample averages correspond to the population estimates since the sample mean is an estimate of the population mean [Cameron and Trivedi, 2005]. In the over-identified case, i.e., when when there are more moment conditions (equations) than parameters, GMM estimation is needed. This is quite often the case in asset pricing given the number of test assets often exceeds the number of parameters (e.g., the concavity parameter of a CRRA utility-based stochastic discount factor).

Following the discussion of Hansen's GMM estimation by Cochrane [2005], begin with the fundamental pricing equation:

$$p_t = E_t \left[m_{t+1} [\mathbf{a}] \, x_{t+1} \right]$$

where **a** represents a vector of parameters $[a_0 \ a_1 \ \cdots a_k]$. The equation can be rewritten in *moment condition* form:

$$E_t \left[m_{t+1}[\mathbf{a}] x_{t+1} - p_t \right] = 0 \tag{9.27}$$

The moment condition is also referred to as the *orthogonality condition*. The expression inside the expectations operator is the pricing error:

$$u_{t+1}[\mathbf{a}] = m_{t+1}[\mathbf{a}] x_{t+1} - p_t \tag{9.28}$$

GMM chooses parameters (**a**) such that the conditional and unconditional mean of the pricing errors (9.28) are zero. GMM arrives at consistent, asymptotically normally, and asymptotically efficient[2] estimates of **a** in a two stage procedure. The first stage obtains utilizes an arbitrary weighting matrix $\mathbf{W}$ (typically $\mathbf{W} = \mathbf{I}$)[3] to obtain a consistent and asymptotically normal parameter vector $\mathbf{a_1}$:

$$\hat{\mathbf{a}}_1 = \text{argmin}_{\{\mathbf{a}\}} g_T \left[\mathbf{a} \right]' \mathbf{W} g_T \left[\mathbf{a} \right]$$

where $g_T[\mathbf{a}]$ represents the sample mean of pricing errors ($u_t[\mathbf{a}]$). An estimate of the sample error variance-covariance matrix is obtained using $\hat{\mathbf{a}}_1$:

$$\hat{\mathbf{S}} = \sum_{j=-\infty}^{\infty} E_t \left[u_t[\hat{\mathbf{a}}_1] u_{t-j}[\hat{\mathbf{a}}_1]' \right]$$

Using $\hat{\mathbf{S}}$, the second stage estimate produces the consistent, asymptotically normal, and asymptotically efficient estimate of **a**, $\hat{\mathbf{a}}_2$:

$$\hat{\mathbf{a}}_2 = \text{argmin}_{\mathbf{a}} g_T[\mathbf{a}]' \hat{\mathbf{S}}^{-1} g_T[\mathbf{a}]$$

The variance-covariance matrix of $\hat{\mathbf{a}}_2$ is:

$$\text{var} \left[\hat{\mathbf{a}}_2 \right] = \frac{1}{T} \left(\mathbf{d}' \hat{\mathbf{S}}^{-1} \mathbf{d} \right)^{-1}$$

---

[2]Cochrane [2005] defines efficient as having "the smallest variance-covariance matrix among all estimators that set different linear combinations of $g_T[\mathbf{a}]$ to zero or all choices of weighting matrix $\mathbf{W}$.

[3]This directs GMM to price all assets equally well. The second stage weighting matrix $\mathbf{W} = \hat{\mathbf{S}}^{-1}$ takes in to account differential variance of asset returns thereby directing GMM to pay less attention to assets with high variances since their sample mean will be a less accurate measure than the population mean.

where

$$\mathbf{d} \equiv \left. \frac{\partial g_T [\mathbf{a}]}{\partial \mathbf{a}} \right|_{\mathbf{a}=\hat{\mathbf{a}}_2}$$

There are several advantages of using GMM estimation in asset pricing scenarios. First, linear and nonlinear asset pricing equations (restrictions) map directly into GMM moment conditions. Second, the estimation allows for serial correlation and non-stationarity (heteroskedasticity) in the pricing errors[4]. Third, the inclusion of instruments, variables in investors information set that are related to future returns or discount factors, is also straightforward. Finally, given that the system of equations for asset pricing tests typically exceeds the number of parameters to be estimated, GMM provides a TJ test-statistic to test if those over-identifying restrictions fit the model. Conditional estimation and the TJ test of over-identifying restrictions are discussed in the following sections.

9.5.1 Conditional vs. unconditional estimation

Begin with the fundamental asset pricing equation:

$$p_t = E_t [m_{t+1} x_{t+1}] \tag{9.29}$$

Let Ω_t represents **all** available time t information therefore the asset pricing equation can be rewritten as:

$$p_t = E [m_{t+1} x_{t+1} | \Omega_t] \tag{9.30}$$

Let I_t represent a subset of **available** information at time t. Recall the law of iterated expectations:

$$E [E [X | \Omega] | I \subset \Omega] = E [X | I]$$

Therefore, taking expectations of both sides of (9.29) conditional on the subset of information I_t:

$$E [p_t | I_t] = E [m_{t+1} x_{t+1} | I_t] \tag{9.31}$$

which implies

$$p_t = E [m_{t+1} x_{t+1} | I_t] \tag{9.32}$$

Therefore an asset can be priced using a subset of all available information. Next consider instrument z_t observed at time t. Multiplying the price and the payoff by z_t:

$$p_t z_t = E_t [m_{t+1} (x_{t+1} z_t)] \tag{9.33}$$

Take unconditional expectations of both sides of (9.33) to obtain:

$$E [p_t z_t] = E [m_{t+1} (x_{t+1} z_t)] \tag{9.34}$$

---

[4]For details see Hansen [1982] and Cochrane [2005]

An intuitive interpretation is as follows. If an investors observes that high values of z_t forecast high returns, the investor might purchase more of the asset at time t. If you consider the price $p = E[p_t z_t]$ and the payoff $x = x_{t+1} z_t$ the pricing equation can be rewritten in unconditional form

$$p = E[mx]$$

Furthermore, Cochrane [2005] shows

$$E[p_t z_t] = E[m_{t+1}(x_{t+1} z_t)] \; \forall z_t \in I_t \quad \Rightarrow \quad p_t = E[m_{t+1} x_{t+1} | I_t] \tag{9.35}$$

Therefore including instruments $z_t \in I_t$, where $I_t \subset \Omega_t$, and taking *unconditional* expectations equivalent to taking *conditional* expectations.

Recall the moment condition for excess returns:

$$E_t[m_{t+1} R_{ei,t+1}] = 0 \tag{9.36}$$

Notice this involves a *conditional* expectation. However, it is often written in *unconditional* form:

$$E[m_{t+1} R_{ei,t+1}] = 0$$

The unconditional form implies the factors of m_{t+1} are constant over time. But what if they are not? A partial solution is to assume coefficients vary with some instrument $z_t \in I_t$. Consider a single factor model with *unconditional* discount factor:

$$m_{t+1} = a + b f_{t+1} \tag{9.37}$$

The *conditional* discount factor can be expressed as:

$$m_{t+1} = a[z_t] + b[z_t] f_{t+1} \tag{9.38}$$

Furthermore, this can conditional discount factor can be expanded to:

$$\begin{aligned} m_{t+1} &= (a_0 + a_1 z_t) + (b_0 + b_1 z_t) f_{t+1} \\ &= a_0 + a_1 z_t + b_0 f_{t+1} + b_1 (z_t f_{t+1}) \end{aligned} \tag{9.39}$$

and now we have replaced the single-factor model with time-varying coefficients (9.37) with a three factor model with fixed coefficients (9.39). Therefore, with fixed coefficients we can used the scaled-factor model with unconditional moment:

$$E[(a_0 + a_1 z_t + b_0 f_{t+1} + b_z (z_t f_{t+1})) R_{i,t+1}] = 1$$

In general, if there are multiple factors and instruments, every factor is multiplied by every instrument (i.e, the Kronecker product):

$$m_{t+1} = b'(f_{t+1} \otimes z_t)$$

9.5.2 TJ test of over-identifying restrictions

As noted in Pynnonen [2007] the reported value of the minimized objective function is the J-statistic:

$$J = g_T \left[\hat{\mathbf{a}}\right]' \hat{\mathbf{S}}^{-1} g_T \left[\hat{\mathbf{a}}\right]$$

Multiplying this by the number of time-series observations T produces the TJ test statistic:

$$TJ \sim \chi^2_{df}$$

where the degrees of freedom, df:

$$df = \#\text{of overidentifying restrictions}$$
$$= \#\text{of moments-}\#\text{of parameters}$$

The null hypothesis for the test statistic is:

$$H_0 : \text{ moment conditions (pricing errors) are zero}$$

Therefore, failure to reject the null is acceptance of the model. To further elaborate on the degrees of freedom, let N represent the number of assets, K the number of factors, L the number of instruments, and P the number of parameters to estimate. As such, the degrees of freedom can be calculated as:

$$df = N(1 + L) - P$$

9.6 Empirical specification of factor pricing models

9.6.1 Identification and excess returns

The mean discount factor is not identified for excess returns. To illustrate:

$$E\left[(a - bf)R_e\right] = 0$$

is not identified while

$$E\left[(1 - bf)R_e\right] = 0$$

is. Why? There is no unique solution to the first equation since:

$$E\left[(a^* - b^* f) R_e\right] = 0$$

where $a^* = 2a$ and $b^* = 2b$ is also satisfied. The normalization $a = 1$ is fine according to Cochrane [2005] who states:

> "lack of identification means precisely that the pricing errors do not depend on the choice of normalization"

However, when forming a linear approximation of the log-utility derived discount factor, choosing $a = 1$ may be inappropriate. Consider the long-run mean real return to capital in the us is 6.5% per annum (1.625% per quarter). A more appropriate choice of a and b can be obtained from the Taylor series expansion of $1/R^W$ about $\bar{R}^W = 1.01625$:

$$f\left[R^W\right] = \frac{1}{R^W}, \qquad f'\left[R^W\right] = -\left(\frac{1}{R^W}\right)^2$$

$$\begin{aligned}
f\left[R^W\right] &\approx f\left[\bar{R}^W\right] + f'\left[\bar{R}^W\right]\left(R^w - \bar{R}^W\right) \\
&= \frac{1}{\bar{R}^W} - \frac{1}{\left(\bar{R}^W\right)^2}\left(R^W - \bar{R}^W\right) \\
&= \frac{2}{\bar{R}^W} - \frac{1}{\left(\bar{R}^W\right)^2}R^W \\
&= \frac{2}{1.0625} - \frac{1}{1.0625^2}R^W \\
&= 1.9680 - 0.9683 R^W
\end{aligned}$$

Similarly

$$E\left[\beta\left(\frac{c_{t+1}}{c_t}\right)^{-\gamma}R_e\right] = 0$$

is not identified while

$$E\left[\left(\frac{c_{t+1}}{c_t}\right)^{-\sigma}R_e\right] = 0$$

However, the mean discount factor *is* identified for real returns:

$$E\left[(a - bf)R_i - 1\right] = 0$$

In this case:

$$E\left[\left(a^* - b^* f\right)R_i - 1\right] \neq E\left[\left(a - bf\right)R_i - 1\right]$$

9.6.2   General SDF moment conditions

9.6.2.1   Unconditional mean excess returns

$$E\left[\frac{u'\left[c_{t+1}\right]}{u'\left[c_t\right]}\left(R_{i,t+1} - R_{f,t+1}\right)\right] = 0 \tag{9.40}$$

Goal To estimate utility function parameters and verify model specification via the TJ test. The estimated utility function parameters are then checked if they fall within plausible ranges. For instance, in the case of CRRA utility, there is a single parameter, γ, the concavity parameter. Large values of γ (>10) are theoretically implausible thereby illustrating the equity premium puzzle is unresolved.

Notes The time rate of preference β is eliminated for identification. It is a constant therefore can eliminated by dividing both sides by β.

9.6.2.2 Conditional mean real returns

$$E\left[\beta\frac{u'[c_{t+1}]}{u'[c_t]}R_{i,t+1}\bigg|Z_t\right] - 1 = 0 \tag{9.41}$$

Goal To estimate utility function parameters and verify model specification via the TJ test. The estimated utility function parameters are then checked if they fall within plausible ranges.

Notes The -1 can be brought inside the expectations operator

9.6.2.3 Conditional mean excess returns

$$E\left[\frac{u'[c_{t+1}]}{u'[c_t]}(R_{i,t+1} - R_{f,t+1})\bigg|Z_t\right] = 0 \tag{9.42}$$

Goal When juxtaposed with (9.41), allows for analysis of the influence of real interest rates on the utility function parameters.

Notes Ferson and Harvey [1992] found (9.41) produced lower concavity parameters (higher intertemporal substitution, i.e., willingness to trade consumption today for consumption tomorrow) compared to (9.42). In other words, *excess* returns imply higher risk aversion than real returns. So what? Investors are more nervous about earning a premium over the risk free rate than the returns themselves? Should the utility function parameter estimates be the same in (9.41) and (9.42)?

9.6.2.4 Conditional mean excess returns with pricing errors

$$E\left[\frac{u'[c_{t+1}]}{u'[c_t]}(R_{i,t+1} - R_{f,t+1} - \lambda_i)\bigg|Z_t\right] = 0 \tag{9.43}$$

Goal To asses the magnitude of mean pricing errors λ_i. Significant λ_i's indicate time variation of conditional expectations is sufficient to reject the model.

Notes Introduction of λ_i's lifts the restriction that unconditional mean *excess* returns are zero.

9.6.3 Linear factor pricing models

9.6.3.1 Single return factor and asset returns

Since the factor is also a return, the asset pricing restriction (9.18) can be applied to the factor:

$$E[f_1] = \lambda_0 + \lambda_1(1) \tag{9.44}$$

Which reveals $\lambda_1 = E[f_1] - \lambda_0$. Substitute λ_1 into (9.18):

$$E[R_i] = \lambda_0 + b_{i1}(E[f_1] - \lambda_0)$$

which is the CAPM if you consider $\lambda_0 = R_f$ and $f_1 = R_m$. Take expectations of the time-series regression (9.19):

$$R_i = b_{i0} + b_{i1}f_{1t} + e_{it}$$
$$E[R_i] = b_{i0} + b_{i1}E[f_1]$$

Next, equate this with the asset pricing restriction (9.18):

$$b_{i0} + b_{i1}E[f_1] = E[R_i] = \lambda_0 + \lambda_1 b_{i1}$$
$$b_{i0} = \lambda_0 + b_{i1}(\lambda_1 - E[f_1]) \tag{9.45}$$

However, this restriction can be simplified by substituting (9.44) into (9.45):

$$b_{i0} = \lambda_0 + b_{i1}(\lambda_1 - (\lambda_0 + \lambda_1)) = (1 - b_{i1})\lambda_0$$

Substituting into the time-series regression (9.19) to obtain the restricted asset pricing model:

$$\boxed{R_{it} = (1 - b_{i1})\lambda_0 + b_{i1}f_{1t} + \epsilon_{it}}$$

As we will see later, this expression is exactly identified when using the factor as the instrument. In other words, there will be two equations and two unknowns (b_{i1} and λ_0). Also, recall that the risk premium for f_1 is $\lambda_1 = E[f_1] - \lambda_0$.

9.6.3.2   Single non-return factor

When the factor is not a return, the system of equations will not be identified (i.e., more parameters than equations). We just saw how a system with factors as returns is identified. Therefore, if we could transform a non-return factor into a return factor we can identify the system.

Cochrane [2005] shows that a factor mimicking portfolio carries the same pricing information as the original factor. There are three possible factor transformations that can be performed.

1. Factor mimicking payoffs constructed from asset payoffs

2. Factor mimicking returns constructed from asset payoffs

3. Factor mimicking excess returns constructed from asset excess returns

We need returns to identify the system so we eliminate option 1. Asset excess returns are more readily accessible than asset payoffs so we shall proceed with option 3. To construct the factor mimicking excess returns regress the true factor on the space of asset excess returns:

$$f_t = \sum_{i=1}^{N} c_i R_{ei,t} + e_{it}$$

Therefore, the predicted values of f_t represent the factor-mimicking excess returns:

$$f_t^* = \sum_{i=1}^{N} \hat{c}_i R_{ei,t}$$

Now use f_t^* in the same fashion as f_t and you arrive at the following restricted asset pricing model:

$$\boxed{R_{it} = (1 - b_{i1}^*) \lambda_0^* + b_{i1}^* f_{1t}^* + \epsilon_{it}}$$

9.6.3.3 Multiple factors when factors are returns

In this case, the asset pricing restriction (9.18) can be incorporated into the time-series regression (9.19) with the restriction:

$$b_{i0} = \lambda_0 + \sum_{k=1}^{K} b_{ik} (\lambda_k - E[f_k])$$

9.6.4 Nonlinear factor models

Consider a two-asset, single instrument case. Technically, the single instrument is actually a constant and the instrument. Therefore the moment conditions are:

$$E\left[\begin{bmatrix} m_{t+1}[\mathbf{a}]R_{1,,t+1} \\ m_{t+1}[\mathbf{a}]R_{2,,t+1} \\ m_{t+1}[\mathbf{a}]R_{1,t+1}z_t \\ m_{t+1}[\mathbf{a}]R_{2,t+1}z_t \end{bmatrix} - \begin{bmatrix} 1 \\ 1 \\ z_t \\ z_t \end{bmatrix}\right] = \begin{bmatrix} 0 \\ 0 \\ 0 \\ 0 \end{bmatrix}$$

Note: The first two rows represent the *unconditional* moment conditions while the last two rows represent the *conditional* moments (since they have scaled prices and payoffs). In general, the moment conditions may be specified as:

$$E\left[(m_{t+1}[\mathbf{a}]R_{i,t+1} - 1) \otimes z_t\right] = 0$$

For example, for CRRA utility:

$$m_{t+1}[\mathbf{a}] = a_0 \left(\frac{c_{t+1}}{c_t}\right)^{-a_1}$$

For log utility:

$$m_{t+1}[\mathbf{a}] = \frac{a_0}{R_{t+1}^W}$$

9.7 Empirical trade-offs

9.7.1 Portfolios vs. Individual securities

Asset pricing models fit portfolios better than individual returns for three major reasons:

1. Portfolio betas are better measured due to lower portfolio volatility and therefore lower residual variance in asset pricing estimations

2. Portfolio betas are more stable over time since individual firm betas vary with conditions affecting the firm's risk level (e.g., size, leverage, specific business environment changes, etc.).

3. Given the high volatility of individual stock returns relative to portfolio returns, you can not reject the hypothesis that average returns are the same for all firms.

9.7.2 Real vs. nominal data

Cochrane [2005] notes that our innocuous asset pricing equation $p = E\,[mx]$ can be based on real or nominal data, provided a consistent choice of prices, returns, and discount factors. For example, if using nominal returns, the discount factor must be based on nominal data.

Nominal values incorporate [expected] inflation:

$$\text{nominal} = \text{real} + E\,[\text{inflation}]$$

whereas real rates exclude inflation. It has been documented that real returns are [negatively] related to inflation [Solnik, 1983, Danthine and Donaldson, 1986]. In other words, real returns can not be fully explained in a model that excludes inflation, which is the case if using strictly real data.

Therefore, given (1) the relevance of inflation, (2) the unavoidable introduction of noise by deflators such as PCE and CPI, and (3) the major source of return data, CRSP, provides nominal returns, the straightforward and parsimonious approach is to use nominal data.

9.7.3 Time-series vs. cross-sectional regression

Often asset pricing tests include multiple firms (N) each with multiple observations (T). Using the CAPM as an example, I show the distinction between time-series and cross-sectional estimations.

CAPM states that expected returns are an increasing function of market risk, with market risk measured by β_i:

$$E\,[R_i] = R_f + \beta_i\,(E\,[R_M] - R_f) \tag{9.46}$$
$$= \lambda_0 + \lambda_m \beta_i$$

Table 9.1: Empirical problems and estimation approaches

Problem	Pooled OLS	F-M	GMM
Serial correlation			$\checkmark$
Heteroscedasticity			$\checkmark$
Cross-sectional correlation		$\checkmark$	$\checkmark$
Generated regressors (β)			$\checkmark$
Time-varying β, λ_0, λ_M		$\checkmark$	

Equation (9.46) represents a <u>cross-sectional</u> regression whose independent variable, β_i, is obtained from the <u>time series</u> regression:

$$R_{i,t} = \alpha_i + \beta_i R_{m,t} + \epsilon_{i,t} \quad \text{or} \qquad (9.47)$$
$$R_{i,t} - R_{f,t} = \beta_i \left(R_{m,t} - R_{f,t} \right) + \epsilon_{i,t}$$

Equations (9.46) and (9.47) expose a couple critical assumptions that are likely to be invalid in practice:

1. α and β are constant throughout time. A firm's market-related risk can and does vary with time since the firm's leverage ratio, management, etc., all vary with time.

2. The market risk premium $\lambda_m = E\left[R_M\right] - R_f$ is time invariant. This assumes the market's degree of risk aversion is constant and we know events can cause the degree of risk aversion to vary (e.g., 9/11).

Nevertheless, there are several approaches to estimating β, λ_0, and λ_m: pooled time-series OLS regression, the Fama-MacBeth procedure, and GMM. Table 9.1 summarizes the ability of the various approaches to address common empirical problems.

Of the procedures listed, only the Fama-MacBeth (F-M) procedure estimates a time-series for the [likely] time-varying β, λ_0, and λ_m. Regarding the pooled OLS approach, Cochrane [2005] states:

> "I emphasize one procedure that is <u>incorrect</u>: pooled time-series and cross-section OLS with no correction of the standard errors. The errors are so highly cross-sectionally correlated in most finance applications that the standard errors so computed are often off by a factor of 10." [emphasis added]

The shortcomings of the Fama-MacBeth procedure can be addressed. Regarding serial correlation and heteroscedasticity when estimating β, (1) OLS is still consistent; (2) there are HAC (heteroscedasticity and autocorrelation corrections) available for the standard errors, and (3) Cochrane [2005] notes:

> "The assumption that the errors are not correlated over time is probably not so bad for asset pricing applications since returns are close to independent."

Similarly, the use of generated regressors (β) in estimating λ_0 and λ_m can be corrected using the Shanken (1992b) correction, although this correction assumes errors from β estimation are i.i.d.

In sum, if one desires to model the time-varying nature of $(\beta, \lambda_0, \lambda_m)$, one could use the F-M procedure and there is good reason to be less concerned about the standard error associated with the mean of the estimates. However, if one is proceeding under the belief that the parameters $(\beta, \lambda_0, \lambda_m)$ are time-invariant, the GMM approach is the simplest method to obtain the most robust standard errors.

9.8   SAS GMM implementation

9.8.1   Nonlinear discount factor model - CCAPM with CRRA utility

Support macro for JT model fit calculation

```
* Macro to compute GMM TJ test statistic p-value;
* Tests H0: Over-identifying restrictions fit the model;
* sometimes the null is stated as:;
* H0: The unconditional and conditional mean of the errors equals zero;
* 2009.10.13 Verified L does NOT include the constant by looking at;
*   the SAS/ETS 'Estimating a Consumption-Based Asset Pricing model';
*   example.;
%macro gmmstats(n, k, l, p);
data est;
    set est;
    format estimate 8.4 stderr 8.4 tvalue 8.2 probt 8.4;
    keep parameter estimate stderr tvalue probt;
run;
data resid;
    set resid;
    format rsquare 8.4 adjrsq 8.4;
    keep equation rsquare adjrsq;
run;
data sumstats;
    set sumstats;
    format pvalue 8.4;
    if _n_ = 2;
    objn = cvalue2;
    N = &n; * number of equations;
    L = &l; * number of instruments excluding constant;
    K = &k; * number of factors;
    P = &p; * number of parameters;
    df = N*(1+L) - P;
    pvalue = 1 - probchi(objn, df);
    keep n k l p df objn pvalue;
run;
title2 "Chi-squared test of over identifying restrictions";
proc print noobs data=sumstats; quit;
title2 "Parameter estimates";
proc print noobs data=est; quit;
%mend;
```

PROC MODEL call

```
ods listing close;
ods output EstSummaryStats=sumstats
            MinSummary=parms
            ResidSummary=resid
            ParameterEstimates=est;
proc model data=modelin;
   exogenous conrat;
   endogenous r11-r13 r21-r23;
   parms beta 1.0 gamma 1.0;
   instruments &instr &instc &instp;
   * define moment conditions;
   eq.s1b1 = beta*conrat**(-gamma)*(1+r11) - 1;
   eq.s1b2 = beta*conrat**(-gamma)*(1+r12) - 1;
   eq.s1b3 = beta*conrat**(-gamma)*(1+r13) - 1;
   eq.s2b1 = beta*conrat**(-gamma)*(1+r21) - 1;
   eq.s2b2 = beta*conrat**(-gamma)*(1+r22) - 1;
   eq.s2b3 = beta*conrat**(-gamma)*(1+r23) - 1;
   fit s1b1 s1b2 s1b3 s2b1 s2b2 s2b3 / itgmm kernel=(parzen,1,0);
quit;
ods listing;
* n=6 equations, k=1 factor (conrat), l=12 instruments, p=2 parameters;
%modelstats(n=6, k=1, l=12, p=2);
```

Sample output   Note, in this case R^2 values are not reported since eq.sibj statements were used.

Chi-squared test of over identifying restrictions						
pvalue	objn	N	L	K	P	df
0.3527	80.0675	6	12	1	2	76

Parameter	Estimate	StdErr	tValue	Probt
beta	0.9666	0.0063	153.41	0.0000
gamma	1.2550	1.0975	1.14	0.2548

Interpretation of output   The p-value of 0.3527 suggest we can accept the null that the moment conditions are zero and the model is well specified. The estimate for β is less than one, consistent with a discount factor. The estimate for γ is consistent with the macroeconomic literature.

9.8.2   Unconditional estimation

Use this when you believe the coefficients of the discount factor parameters are constant over time.

9.8.2.1 Using returns

For a 1-factor model, the moment conditions are:

$$E\left[R_{it}\left(a_0 + a_1 f_{1t}\right) - 1\right] = 0$$
$$E\left[R_{i,t} - \left(b_{i0} + b_{i1} f_{1t}\right)\right] = 0$$
$$E\left[R_{i,t} - \left(\lambda_0 + \lambda_1 b_{i1}\right)\right] = 0$$

The first equation tests whether or not the discount factor $m_t = a_0 + a_1 f_{1t}$ can be used to price the set of assets. The second equation is the time series regression to estimate b_{ik}. The third equation is the asset pricing restriction.

9.8.2.2 Using excess returns

For a 1-factor model with excess returns for test assets and factors, the moment conditions are:

$$E\left[R_{ei,t}\left(a_0 + a_1 f_{1t}\right)\right] = 0$$
$$E\left[R_{ei,t} - \left(b_{i0} + b_{i1} f_{1t}\right)\right] = 0$$
$$E\left[R_{ei,t} - \lambda_1 b_{i1}\right] = 0$$

The first equation tests whether or not the discount factor $m_t = a_0 + a_1 f_{1t}$ can be used to price the set of assets. The second equation is used to estimate b_{i1}. The fourth equation is the asset pricing restriction. Note, when using excess returns test assets and factors, $\lambda_0 = 0$.

9.8.3 Conditional estimation

Use this when you believe the coefficients of the discount factor parameters vary over time and you have chosen instruments (variables) that impact the value of those coefficients over time.

Chapter 10

GRS Test Statstic for Factor Model Fit

10.1   What is the GRS test statistic for?

Begin with the following asset pricing equations:

$$m_{t+1} = a_0 + a_1 f_{1,t+1} + a_2 f_{2,t+1} + \cdots + a_K f_{k,t+1} \tag{10.1}$$
$$E_T[R_{ei}] = \lambda_0 + \lambda_1 b_{i1} + \lambda_2 b_{i2} + \cdots + \lambda_K b_{iK} \tag{10.2}$$
$$R_{ei,t} = b_{i0} + b_{i1} f_{1,t} + b_{i2} f_{2,t} + \cdots + b_{iK} f_{K,t} + e_{i,t} \tag{10.3}$$

where

$$m_{t+1} = \text{stochastic discount factor from } p = E[mx]$$
$$\lambda_k = \text{market price of risk for factor } k$$
$$R_{ei} = \text{excess return for asset } i$$
$$f_k = \text{factor believed to be related to } m \text{ (e.g., mkt, smb, hml)}$$

The GRS test statistic [Gibbons et al., 1989], W, tests the null hypothesis.

$$H_0: \ b_{i0} = 0 \quad \forall i$$

which should *not* be rejected if the factors completely explain excess returns. Note that the GRS test relies on the time-series regression (10.3) and does not involve the cross-sectional regression (10.2).

The GRS test statistic (W) has better small sample properties than the Wald, Lagrange Multiplier, and Likelihood ratio tests [Gibbons et al., 1989]. Specifically, it is less likely to falsely reject the null (intercepts are all zero).

10.2　Estimation of W

Let N represent the number of equations (assets, portfolios) estimated, T the number of observations per asset, and K the number of factors in the model (e.g., 3 for the Fama and French [1993] 3 factor model).

10.2.1　Single factor model

When there is a single factor $(K = 1)$:

$$W = \hat{b}_0' \hat{\Sigma}^{-1} \hat{b}_0 / \left(1 + \hat{\theta}_1^2\right)$$

where

$$\hat{b}_0 = \text{column vector of intercept estimates } (N \times 1)$$
$$\hat{\Sigma} = \text{unbiased resuidual covariance matrix } (N \times N)$$
$$\hat{\theta}_1 = \frac{\bar{r}_1}{s_1} = \text{sample mean of } f_1 \text{ divided by the sample variance } (1 \times 1)$$

To compute the F-statistic:

$$\frac{T(T - N - 1)}{N(T - 2)} W \sim F_{df1, df2} \tag{10.4}$$

where $df1 = N$ and $df2 = T - N - 1$. Note, under the null,(10.5) refers to a central F distribution, the technique employed by Fama and French [1993].

10.2.2　Multi factor model

When there are multiple factors $(K > 1)$:

$$W = \left(1 + \bar{f}' \hat{\Omega}^{-1} \bar{f}\right)^{-1} \hat{b}_0' \hat{\Sigma}^{-1} \hat{b}_0$$

where

$$\bar{f} = \text{column vector of sample means } (K \times 1)$$
$$\hat{\Omega} = \text{sample variance-covaraince matrix for } f \ (K \times K)$$
$$\hat{b}_0 = \text{column vector of intercept estimates } (N \times 1)$$
$$\hat{\Sigma} = \text{unbiased resuidual covariance matrix } (N \times N)$$

To compute the F-statistic:

$$\left(\frac{T}{N}\right)\left(\frac{T - N - K}{T - K - 1}\right) W \sim F_{df1, df2} \tag{10.5}$$

where $df1 = N$ and $df2 = T - N - K$. Note, under the null, (10.5) refers to a central F distribution, and this is the technique employed by Fama and French [1993].

10.3 SAS code

10.3.1 Obtain data

Data on the 25 portfolios sorted on size and book-to-market equity are obtained from WRDS and merged with the Fama and French factors, also from WRDS, using the following code:

```
* ----------------------------------;
* macro to generate return variables;
* ----------------------------------;
%macro ff25;
    %do i=1 %to 5;
        %do j = 1 %to 5;
        r&i&j = s&i.b&j._vwret - rf;
        %end;
    %end;
%mend;
    * ----------------------------------;
    * macro to generate moment conditions;
    * ----------------------------------;
%macro ffeq25;
    %do i = 1 %to 5;
        %do j = 1 %to 5;
        eq.s&i.b&j = r&i&j - (a&i&j + b&i&j*mkt + s&i&j*smb +h&i&j*hml);
        %end;
    %end;
%mend;
    * --------------------------;
    * read in fama french factors;
    * --------------------------;
data ff3;
    set ff.factors_monthly;
    date = dateff;
    mkt = mktrf;
    keep date mkt smb hml rf umd;
run;
    * ---------------------------;
    * merge with portfolio returns;
    * ---------------------------;
%merge2(ff3, ff.portfolios25, monthly, date);
    * ---------------------;
    * compute excess returns;
    * ---------------------;
data monthly;
    set monthly;
    %ff25;
    keep date mkt smb hml rf umd r11-r15 r21-r25 r31-r35 r41-r45 r51-r55;
    yyyymm = 100*year(date) + month(date);
    if 196307<= yyyymm <= 199112;
    drop yyyymm;
run;
```

10.3.2   Parameter estimation

The coefficients are estimated via a simultaneous regression of all 25 portfolios using PROC MODEL and the GMM estimation technique with the following code:

```
* --------------------------;
* perform asset pricing tests;
* --------------------------;
title "Fama French 3-factor model";
ods listing close;
ods output parameterEstimates=est
           testResults=test;
proc model data=monthly;
   * define parameters to be estimated;
   parms a11-a15 a21-a25 a31-a35 a41-a45 a51-a55
         b11-b15 b21-b25 b31-b35 b41-b45 b51-b55
         s11-s15 s21-s25 s31-s35 s41-s45 s51-s55
         h11-h15 h21-h25 h31-h35 h41-h45 h51-h55;
   * define instruments;
   instruments mkt smb hml;
   * define moment conditions;
   %ffeq25;
   * fit the model;
   fit s1b1 s1b2 s1b3 s1b4 s1b5
       s2b1 s2b2 s2b3 s2b4 s2b5
       s3b1 s3b2 s3b3 s3b4 s3b5
       s4b1 s4b2 s4b3 s4b4 s4b5
       s5b1 s5b2 s5b3 s5b4 s5b5 / gmm kernel=bart
                                 outest=est1 outs=sigma;
   * test intercepts;
   test a11, a12, a13, a14, a15,
        a21, a22, a23, a24, a25,
        a31, a32, a33, a34, a35,
        a41, a42, a43, a44, a45,
        a51, a52, a53, a54, a55, /Wald;
run;
quit;
```

Note: that QUIT at the end is pretty important. Otherwise, the next ODS output statement will generate warnings.

10.3.3   GRS test statistic estimation

Through the use of PROC CORR and PROC IML, the GRS test statistic is calculated with the following code:

```
* -----------------------------------;
* compute GRS test statistic and p-value;
* -----------------------------------;
* compute variance-covariance matrix;
ods output cov=omega simpleStats=stats;
```

```
proc corr nocorr cov data=monthly;
  var mkt smb hml;
run;
quit;
* compute statistic;
ods listing;
title2 "GRS test statistic and P-value";
proc iml;
  * factor sample variance-covariance matrix;
  use omega;
  read all var {mkt smb hml} into omegah;
  * residual sample variance-covariance matrix;
  * note: sigma comes from the OUTS= option of the FIT;
  *       statement in PROC MODEL;
  use sigma;
  read all var {s1b1 s1b2 s1b3 s1b4 s1b5
                s2b1 s2b2 s2b3 s2b4 s2b5
                s3b1 s3b2 s3b3 s3b4 s3b5
                s4b1 s4b2 s4b3 s4b4 s4b5
                s5b1 s5b2 s5b3 s5b4 s5b5} into sigmah;
  * factor means;
  use stats;
  read all var {mean} into fbar;
  * intercept estimates and number of observations;
  use est1;
  read all var {a11 a12 a13 a14 a15
                a21 a22 a23 a24 a25
                a31 a32 a33 a34 a35
                a41 a42 a43 a44 a45
                a51 a52 a53 a54 a55} into b0;
  read all var {_NUSED_} into t;
  * N = number of assets (portfolios);
  * K = number of factors (e.g., 3 for MKT, SMB, and HML);
  * T = number of observations;
  N = 25;
  K = 3;
  iomega = inv(omegah);
  isigma = inv(sigmah);
  W = ((1 + fbar'*iomega*fbar)**-1)*b0*isigma*b0';
  F = (T/N)*((T-N-K)/(T-K-1))*W;
  df1 = N;
  df2 = T-N-K;
  pvalue = 1 - probf(F,df1,df2);
  print F pvalue;
quit;
```

10.4   Conclusion: comparison between GRS and Wald test statistics

Running the aforementioned code produces the following:

Method	Test statistic	p-value
GRS	1.6771162	0.024259
Wald	47.62	0.0041

As the table indicates, the Wald test statistic is rejects the null (all intercepts are zero) at the 1% level whereas the GRS statistic does not. For the same sample period, portfolios, and 3 factor model, Fama and French [1993] obtained a p-value of 0.039. The authors state:

> "Despite its marginal rejection in the F-tests, our view is that the three-factor model does a good job on the cross section of average stock returns."

The authors also mention individual tests of coefficients find that only 1 out of 25 intercepts is "much different from zero." I find 6 out of 25 are statistically different from zero at the 5% level. Similarly, Chordia and Shivakumar [2006] obtain a GRS p-value of 0.04 and conclude their model "explains" momentum. In sum, obtaining a GRS p-value of greater than 0.01 appears to be sufficient to make a claim of a model's explanatory ability.

Chapter 11

Time series econometrics

The information in this chapter flosely follows Enders [2004] and the reader is encouraged to reference Enders' text for further details.

11.1 Stationarity

11.1.1 Background

A stochastic process is *covariance stationary* when it has time-invariant and finite mean, variance, and covariances:

$$E[y_t] = E\left[y_{t-s}\right] = \mu \quad \forall s$$

$$E\left[(y_t - \mu)^2\right] = E\left[(y_{t-s} - \mu)^2\right] = \sigma_y^2 \quad \forall s$$

$$E[(y_t - \mu)(y_{t-s} - \mu)] = E[(y_{t-j} - \mu)(y_{t-j-s} - \mu)] = \gamma_s \quad \forall s \text{ and any } j$$

A process is said to be *strict stationary* if the PDF (which encompasses higher order moments) over sub-samples is time-invariant.

Difference stationary The following random walk process is non-stationary (has a unit root):

$$y_t = y_{t-1} + \epsilon_t$$

However, the first difference is stationary:

$$\Delta y_t \equiv y_t - y_{t-1} = \epsilon_t$$

as such, the process y_t is said to be *difference stationary* or *integrated of order 1* (I[1]). Next, consider the random walk plus drift model

$$y_t = a_0 + y_{t-1} + \epsilon_i$$

Given the initial condition y_0 the general solution for y_t is:

$$y_t = y_0 + a_0 t + \sum_{i=1}^{t} \epsilon_i$$

which is non-stationary. Taking the first difference yields $\Delta y_t = a_0 + \epsilon_t$, a stationary process.

Trend stationary A process is said to be trend stationary if it can be made stationary by detrending. Consider the following process with a deterministic trend.

$$y_t = a_0 + bt + \epsilon_t$$

Removal of bt yields the stationary process $y_t = a_0 + \epsilon_t$.

Consider $\{y_t\}$ generated from a random walk (unit root process) and regressed using the following equation:

$$y_t = a_1 y_{t-1} + \epsilon_t \tag{11.1}$$

Presume the OLS regression resulted in $\widehat{a}_1 = 0.96$ with a standard error of 0.020. With OLS, the null hypothesis is $H_0 : a_1 = 0$, and the highly significant t-statistic indicates rejection of the null. Therefore, all we know is $a_1 \neq 0$ and we have not tested for a unit root ($a_1 = 1$). By subtracting y_{t-1} from both sides of (11.1), we are able to obtain an equivalent regression that enables OLS to test for $a_1 = 1$. In addition, drift and deterministic trends may be included resulting in the following three expressions.

$$\Delta y_t = \gamma y_{t-1} + \epsilon_t \tag{11.2}$$
$$\Delta y_t = a_0 + \gamma y_{t-1} + \epsilon_t \tag{11.3}$$
$$\Delta y_t = a_0 + \gamma y_{t-1} + a_2 t + \epsilon_t \tag{11.4}$$

The associated test statistics, null hypotheses, and alternative hypotheses of these equations are summarized in the following table:

H_a (stationary)	H_0 (non stationary)	test statistic
zero mean: $\Delta y_t = \gamma y_{t-1} + \epsilon_t$	$\Delta y_t = \epsilon_t$	τ
single mean: $\Delta y_t = a_0 + \gamma y_{t-1} + \epsilon_t$	$\Delta y_t = a_0 + \epsilon_t$	τ_μ
	$\Delta y_t = \epsilon_t$	ϕ_1
trend: $\Delta y_t = a_0 + \gamma y_{t-1} + a_2 t + \epsilon_t$	$\Delta y_t = a_0 + a_2 t + \epsilon_t$	τ_τ
	$\Delta y_t = a_0 + \epsilon_t$	ϕ_3
	$\Delta y_t = \epsilon_t$	ϕ_2

Recall rejection of the null is acceptance of the alternative. Alternatively, failure to reject the null implies non-stationarity. The interpretation of test statistic significance, p-values, and hypotheses are summarized in the following table:

statistic	p-value	null	process is
significant	low	reject	stationary
insignificant	high	accept	non-stationary

11.1.2 Causes and consequences of non-stationary time series

Non-stationary time series result from random walk (unit root) or processes with a trend (deterministic or stochastic). Consider the former with the following set of equations:

$$y_t = a_0 + b_0 z_t + \epsilon_t \tag{11.5}$$

$$y_t = y_{t-1} + \epsilon_{yt}$$

$$z_t = z_{t-1} + \epsilon_{zt}$$

That meaningfulness of regression (11.5) is dependent on stationarity of $\{y_t\}$ and $\{z_t\}$. The following five cases are described in Enders [2004]:

$\{y_t\}$	$\{z_t\}$	$\{e_t\}$	Regression
stationary (I[0])	stationary (I[0])	stationary	appropriate
non-stationary ($I[d_1]$)	non-stationary ($I[d_2]$)	non-stationary	meaningless
trend stationary	difference stationary ($I[d_2]$)	non-stationary	meaningless
non-stationary ($I[d^*]$)	non-stationary ($I[d^*]$)	non-stationary	meaningless
non-stationary ($I[d^{**}]$)	non-stationary ($I[d^{**}]$)	stationary	appropriate

Note: In the final scenario (integrated of same order yet error term is stationary), the variables are said to be *cointegrated* which implies a common stochastic trend.

11.1.3 Detection - dftest macro

The Augmented Dickey-Fuller stationarity test can be performed with DFTEST macro:

```
* Standard ADF test;
%DFTEST(datain, varname, dlag=1, trend=2, outstat=adf);
%put &dftest;
* Seasonal ADF test (quarterly data);
%DFTEST(datain, varname, dlag=4, trend=1, outstat=adf);
%put &dftest;
```

Where

- DATAIN is the input data set.

- VARNAME is the variable to test.

- DIFF represents the differencing to be applied to VARNAME prior to testing.

- DLAG is (1, 2, 4, 12) for non-seasonal, semi-annual, quarterly, or monthly unit root testing.

- TREND is (0, 1, 2) for the model with no drift/trend, model with drift, and model with drift and trend. note, a TREND value of 2 applies only when DLAG is 1.

- The &DFTEST environment variable is the p-value of the τ test statistic.

- OUTSTAT writes the test statistic, parameter estimates, and other statistics to an output data set.

- There is currently no way to get ϕ test statistics using the DFTEST macro.

11.1.4   Detection - proc arima

Alternatively, proc arima can be used to perform the Dickey-Fuller stationarity test:

```
%let adf1 = (adf=4 dlag=1);  * basic unit root test;
%let adf4 = (adf=4 dlag=4);  * seasonal unit root test for;
                             * quarterly data;
%let pp4 = (phillips=4);
%macro adftest(x, y);
proc arima data=&x;
   identify var = &y nlag=4 stationarity = &adf1;
   identify var = &y(1) nlag=4 stationarity = &adf1;
run;
%mend;
```

where

- ADF represents the lagged differences to use in the model.

- DLAG is (1, 2, 4, 12) for non-seasonal, semi-annual, quarterly, or monthly unit root testing.

- The null hypothesis is H_0: d.g.p. contains a unit root.

Consider the sample output of this procedure:

```
            Dickey-Fuller Unit Root Tests
Type         Lags      Rho Pr < Rho     Tau Pr < Tau      F Pr > F
Zero Mean      0 -27.6594   <.0001    -3.97   0.0001
Single Mean    0 -124.505   0.0001   -11.64   <.0001   67.73 0.0010
Trend          0 -125.982   0.0001   -11.72   <.0001   68.67 0.0010
```

The low p-values for all τ statistics indicates rejection of the null therefore the process is stationary.

11.1.5   Detection - Phillips-Perron tests

The augmentation (addition of lagged difference terms) in the ADF (augmented dickey-fuller) tests is to account for possible serial correlation in the error terms. In addition, the ADF test assumes the error terms are i.i.d. The Phillips-Perron (PP) test use nonparametric (no distributional assumptions of error terms) statistical methods that account for serial corelation in error terms.

```
proc arima data=x;
   identify var = y stationarity = (pp=4);
run;
```

For more information see the SAS documentation.

11.1.6 Detection - Johansen rank test for cointegration

If two time series, $\{x_t\}$ and $\{y_t\}$ are integrated of the same order, regression of y_t onto x_t may still be appropriate if they are cointegrated. To simultaneously test for a unit root (ADF test) and cointegration, use the following code.

```
%let alpha = 0.05;
proc varmax data=coidata;
    model x y / p=4 dftest cointtest=(johansen siglevel=&alpha);
run;
```

The first output table of this code is the ADF test is shown in the figures to follow.

Figure 11.1: Dickey-Fuller Unit Root Tests

Variable	Type	Rho	Pr < Rho	Tau	Pr < Tau
x	Zero Mean	0.07	0.6921	3.05	0.9990
	Single Mean	-1.70	0.8052	-1.57	0.4857
	Trend	-13.58	0.1733	-3.43	0.0642
y	Zero Mean	0.08	0.6948	2.82	0.9983
	Single Mean	0.53	0.9743	0.44	0.9816
	Trend	-6.17	0.7009	-1.43	0.8317

Figure (11.1) indicates both series have a unit root since we are unable to reject the null of the presence of a unit root (although the presence of a unit root could be rejected for Y at the 10% level). Next is the cointegration rank test, shown in Figure (11.2).

Figure 11.2: Cointegration Rank Test Using Trace

2A: Cointegration Rank Test Using Trace

H0: Rank=r	H1: Rank>r	Eigenvalue	Trace	5% Critical Value	Drift in ECM	Drift in Process
0	0	0.3798	18.5200	15.34	Constant	Linear
1	1	0.0961	3.2316	3.84		

2B: Cointegration Rank Test Using Trace Under Restriction

H0: Rank=r	H1: Rank>r	Eigenvalue	Trace	5% Critical Value	Drift in ECM	Drift in Process
0	0	0.5167	28.8514	19.99	Constant	Constant
1	1	0.1602	5.5869	9.13		

Figure (11.2) revelas the trace test statistic lies within the "reject" region for H_0 : rank=0 and the "accept" region for H_0 : rank=1 in both cases. So which model is correct? Figure (11.3) shows the hypothesis test of the restriction, i.e., which case fits the time series, case A (no restriction) or case B (under restriction).

Figure 11.3: Case determination

```
                  Hypothesis of the Restriction

                          Drift          Drift in
           Hypothesis     in ECM         Process
           H0             Constant       Constant
           H1             Constant       Linear
               Hypothesis Test of the Restriction

                    Restricted
  Rank  Eigenvalue  Eigenvalue   DF   Chi-Square   Pr > ChiSq
   0      0.3798      0.5167       2      10.33        0.0057
   1      0.0961      0.1602       1       2.36        0.1249
```

To interpret Figure (11.3), begin by looking at the row for rank=1 (we previously determined the rank is 1). The p—value of 0.1249 indicates we can not reject the null that the restircited model, case B, is the better fit for the data. Therefore we an conclude (1) variables X and Y have a unit root (failure to reject H_0 : d.g.p. contains a unit root), (2) X and Y are cointegrated of order 1, and (3) the model is of the form of Case B (constant drift in ECM, constant drift in process).

11.1.7   Correction

Differencing   Consider random walk model:

$$y_t = y_{t-1} + \epsilon_t$$

The first difference is: $\Delta y_t = \epsilon_t$ is stationary since $E[\epsilon_t \epsilon_{t-s}] = 0$ for all s. In other words, prior shocks have no persistent effects.

Detrending   The procedure to remove a deterministic trend is as follows:

1. Begin with polynomial time-series trend:

$$y_t = a_0 + \sum_{i=1}^{n} a_i t^i + e_t$$

where e_t is a stationary process (could be ARMA).

2. Determine the appropriate value of n by regressing:

$$y_t = a_0 + \sum_{i=1}^{n} a_i t^i + \epsilon_t$$

where ϵ_t is a white noise process. Succesively reduce n until a nonzero coefficient is obtained.

3. Compute $\widehat{y}_t$

4. The series $y_t - \widehat{y}_t$ is now stationary and can be estimated using traditional methods (such as ARMA).

11.2 Autocorrelated error terms

$$u_t = \sum_{i=1}^{m} \rho_i u_{t-i} + \epsilon_t \tag{11.6}$$

11.2.1 Causes and consequences

Autocorrelated error terms have the following possible causes:

1. Incorrect specificationor functional form.

2. Ommitted variables having a time trend.

3. Transformation of variables.

4. Effect of random disturbance lasts more than 1 period (e.g., a machine breaks down).

Naturally, there are consequences to having autocorrelated errors:

1. Residual variance is likely to underestimate the true variance.

2. Underestimated residual variance leads to overestimated R^2.

3. Underestimated coefficient standard errors leads to overestimated t values.

4. t and F tests are no longer valid.

11.2.2   Detection

Autocorrelated errors can be detected using the Durbin-Watson test statistic as shown in the following code:

```
proc autoreg data=a;
   model y = x / dw=m dwprob;
run;
```

Where m represents the order of autocorrelation to look for as in equation (11.6) and DWPROB tells sas to output p-values. Use $m = 4$ for quarterly data and $m = 12$ for monthly data. Let $p = \Pr\left[dw < \widehat{dw}\right]$ represent the probability that the true test statistic is less than the estimated test statistic. The null hypothesis is

$$H_0 : \quad \text{no autocorrelation}$$

As such, low p-values indicate rejection of the null and therefore autocorrelation is present. Note, this test should not be used to determine the order, m, of the autoregressive errors.

11.2.3   Correction

To determine m, the BACKSTEP option can be used:

```
proc autoreg data=a;
   model y = x /method=ml nlag=n backstep dwprob;
run;
```

Where n is the highest-order lag to test for autocorrelation. In general, it should be set to 5 for quarterly data and 13 for monthly data but higher values may be used. Sample output from this procedure with $m = 20$:

```
Backward Elimination of
Autoregressive Terms
Lag     Estimate    t Value    Pr > |t|
19     -0.013233     -0.13      0.8955
18      0.016053      0.17      0.8659
 3      0.026320      0.27      0.7908
14     -0.030322     -0.31      0.7604
15     -0.047745     -0.53      0.5996
11      0.041524      0.42      0.6748
 7     -0.040770     -0.53      0.5937
16     -0.091846     -0.88      0.3794
17      0.066219      0.79      0.4327
 5      0.101057      0.96      0.3413
 8     -0.131968     -1.27      0.2071
 2     -0.096406     -1.15      0.2519
10     -0.127248     -1.44      0.1509
20     -0.080096     -1.39      0.1674
12      0.099733      1.45      0.1490
```

Estimates of Autoregressive Parameters

Lag	Coefficient	Standard Error	t Value
1	-0.452260	0.061459	-7.36
4	-0.657737	0.060928	-10.80
6	0.165533	0.061060	2.71
9	0.290591	0.085522	3.40
13	-0.167386	0.073611	-2.27

This output indicates lags 1, 4, 6, 9, and 13 are to be included in the autoregressive specification of error terms. As such, estimation of the model should be performed with the NLAG = (1 4 6 9 13) option.

11.3   Heteroscedastic error terms

$$E\left[u_i^2\right] = \sigma_i^2 \neq \sigma^2$$

11.3.1   Causes and consequences

Heteroscedastic errors have the following possible causes:

1. Using average data. E.g., average consumption per state of population n_i, average income per state, both of which lead to an average disturbance per state.

2. Using functions of the form:
$$\text{sales per store} = f\left[\text{advertising,expenditures,income}\right]$$

3. Error-learning models: as people learn they reduce their errors.

4. Increase in variance as income increases.

5. Improved data collection coud lead to reduced σ_i^2 over time.

6. The presence of outliers.

7. Model mis-specification or omitted variables.

8. Skewnewss in regressor distribution.

9. Incorrect data transformation and incorrect functional form (linear vs. log-linear).

Naturally, there are consequences to having autocorrelated errors:

1. OLS estimators are still unbiased but...

2. OLS estimators are not BLUE. E.g., $var\left[\widehat{\beta}_{OLS}\right] \geq var\left[\widehat{\beta}_{GLS}\right]$.

3. OLS estimate of σ^2 is biased.

4. t tests are invalid due to (2) and (3).

11.3.2   Detection - proc autoreg

Heteroscedastic errors can be detected using the ARCHTEST option in the AUTOREG procedure:

```
proc autoreg data=a;
    model y = x / nlag = (1 4 6 9 13) archtest;
run;
```

Let $p_q = \Pr\left[Q > \widehat{Q}\right]$ and $p_{lm} = \Pr\left[LM > \widehat{LM}\right]$ represent the p-values for Q and Lagrange Multiplier tests, respectively. The null hypothesis is:

$$H_0: \quad \text{no heteroscedasticity}$$

As before, the null is rejected whe p-values are low. Note nlag = (1 4 6 9 13) specifies the error term autoregressive process as identified in Section 11.2.3.

11.3.3   Detection - proc reg

To detect heteroscedasticity using PROC REG, use SPEC option. The null hypothesis is the error terms are homoscedastic.

```
proc reg data=datain;
    model y = k / spec;
run;
```

Which produces the following output:

```
            The REG Procedure
               Model: cd
          Dependent Variable: y
          Test of First and Second
             Moment Specification
      DF     Chi-Square     Pr > ChiSq
       2          9.57         0.0084
```

In this case the low p-value indicates rejection of the null. Therefore error terms are to be considered heteroscedastic.

11.4   Simultaneous heteroscedasticity and autocorrelation

Sayrs (1989), Pooled Time Series Analysis, suggests:

"Autoregression can only be detected after heteroscedasticity is controlled for."

However, there are tests to handle heteroscedasticity and autocorrelation simultaneously (Bera and Jarque 1981).

11.4.1   Approach 1 - correct for both (Newey-West HAC)

The simple way:

```
* correct for autocorrelation and heteroscedasticity;
ods listing close;
ods output parameterestimates=nw residsummary=fit;
proc model data=mylib.qtrlog;
   parms a0 a1;
   y = a0 + a1*k;
   fit y / gmm maxiter=500 kernel=bart;
run;
ods listing;
proc print data=nw;
   id parameter;
   format estimate stderr 7.4;
run;
data fit;
   set fit;
   keep esttype equation rsquare adjrsq;
run;
proc print data=fit;
   id equation;
run;
```

Which produces the output:

```
                        The SAS System
    10:22 Friday, November 9, 2007
    Parameter   EstType    Estimate    StdErr    tValue    Probt      DF
       a0       GMM Esti     1.8594    0.5864      3.17    0.0019   118.0
       a1       GMM Esti     0.3503    0.1067      3.28    0.0013   118.0
                        The SAS System                                 2
                                   10:22 Friday, November 9, 2007
            Equation    EstType    RSquare    AdjRSq
               y        GMM Esti    0.1391     0.1318
```

Fama MacBeth regressions and Newey West correction

Running a Fama-Macbeth regression in SAS is quite easy, and doesn't require any special macros. The following code will run cross-sectional regressions by year for all firms and report the means.

```
* Step 1 - generate a time series of parameter estimates;
ods listing close;
ods output parameterestimates=pe;
* Note: dset contains monthly observations;
proc reg data=dset;
   by year;
   model depvar = indvars;
run;
quit;
ods listing;
```

```
    proc means data=pe mean std t probt;
       var estimate;
       class variable;
    run;
```

Since the results from this approach give a time-series, it is common practice to use the Newey-West adjustment for standard errors. This can be done in SAS as follows:

```
    * Step 2 - compute mean and Newey-West adjusted errors;
    proc sort data=pe;
       by variable;
    run;
    %let lags=3;
    ods output parameterestimates=nw;
    ods listing close;
    proc model data=pe;
       by variable;
       instruments / intonly;
       estimate=a;
       fit estimate / gmm kernel=(bart,%eval(&lags+1),0);
    run;
    quit;
    ods listing;
    proc print data=nw;
       id variable;
       var estimate--df;
       format estimate stderr 7.4;
    run;
```

Note that the lag length is set by defining a macro variable, LAGS. The approach here is to use GMM to regress the time-series estimates on a constant, which is equivalent to taking a mean. This works because the Newey-West adjustment gives the same variance as the GMM procedure. (See Cochrane [2005] for details.)

11.4.2   Approach 2 - Model both (AR(m)-GARCH)

With unconditional GARCH, the GARCH process is a function of past residuals and past conditional variances. In the unconditional GARCH, the GARCH process can be a function of exogenous variables.

$$y_t = \mathbf{x}'\beta + u_t$$

$$u_t = \sum_{i=1}^{m} \rho_i u_{t-i} + \epsilon_t$$

$$\epsilon_t = \sqrt{h_t} e_t$$

$$h_t = \omega + \sum_{i=1}^{q} \alpha_i \epsilon_{t-i}^2 + \sum_{j=1}^{p} \gamma_j h_{t-j} \qquad (11.7)$$

$$e_t \sim IN(0,1)$$

The output of the heteroscedasticity detection code in Section 11.3.2 is as follows:

```
         Q and LM Tests for ARCH Disturbances
Order           Q     Pr > Q         LM     Pr > LM
  1        5.6442     0.0175     5.5969      0.0180
  2        5.6922     0.0581     6.0686      0.0481
  3        9.0224     0.0290     9.8494      0.0199
  4       57.1468     <.0001    47.2399      <.0001
  5       61.6711     <.0001    47.2556      <.0001
  6       62.1021     <.0001    47.9185      <.0001
  7       62.2097     <.0001    49.1957      <.0001
  8       70.6225     <.0001    51.1148      <.0001
  9       71.1914     <.0001    51.8604      <.0001
 10       71.5842     <.0001    51.8856      <.0001
 11       71.6500     <.0001    51.8859      <.0001
 12       71.6630     <.0001    53.5570      <.0001
```

Significant lagrange mutipliers at all lags are suggestive of a long memory process ($p > 0$). A long memory process can be captured can be captured with $p = 1$ (see equation 11.7). The following code fits an AR(1 4 6 9 13)-GARCH(1,1) model to the data:

```
proc autoreg data=a;
   model y = x / nlag = (1 4 6 9 13) garch = (q=1,p=1) maxit=50;
run;
```

The output is as follows:

```
                          GARCH Estimates
SSE                4.37042948   Observations              140
MSE                   0.03122   Uncond Var         0.02323037
Log Likelihood     103.377871   Total R-Square         0.8820
SBC                -157.33932   AIC                -186.75574
Normality Test         2.5925   Pr > ChiSq             0.2736
                                     Standard              Approx
Variable      DF      Estimate         Error   t Value    Pr > |t|
Intercept      1       -1.0548        0.3225     -3.27      0.0011
lnk            1        0.8030        0.0599     13.40      <.0001
```

AR1	1	-0.4759	0.0615	-7.74	<.0001
AR4	1	-0.7072	0.0699	-10.12	<.0001
AR6	1	0.0645	0.0692	0.93	0.3507
AR9	1	0.3448	0.0950	3.63	0.0003
AR13	1	-0.2280	0.0563	-4.05	<.0001
ARCH0	1	0.001203	0.000698	1.72	0.0846
ARCH1	1	0.3288	0.1332	2.47	0.0135
GARCH1	1	0.6194	0.1296	4.78	<.0001

The normality test statistic applies to the following null hypothesis:

$$H_0 : \quad \text{GARCH model residuals, } e_t, \text{ are normally distributed}$$

as before, low p-values indicate a significant test statistic and rejection of the null. In this case, the test statistic is insignificant therefore the null cannot be rejected and the GARCH residuals terms are presumed to be normally distributed.

Conditional GARCH   There are also several conditional GARCH techniques available: EGARCH, IGARCH, GARCH-in-mean, etc. Perhaps in a later edition of this book. However, I would mention the principle of parsimony likely applies here as well.

11.4.3   Aporoach 3 - Model AR(m), correct for heteroscedasticity

If you like the hybrid concept you may try the following untested code.

```
* model autocorrelation, correct for heteroscedasticity;
proc model data=mylib.qtrlog;
    parms a0 a1 ;    y = a0 + a1*k;
    %ar(mu, 5, y, 1 3 4 5);
    fit y / gmm maxiter=500;
    instruments k t k2 t2 i c;
run;
```

Chapter 12

COMPUSTAT on WRDS

12.1 Working with dates

12.1.1 Annual data

Data item	format	Description
fyr	2 digit integer	fiscal year end moth
date	4 digit integer	fiscal year
yeara	sas date	fiscal year

If $fyr \leq 5$, COMPUSTAT sets $yeara$ equal to the beginning of fiscal year (say year t) even though the annual report appears in year $t + 1$. If $fyr \geq 6$, compustat sets $yeara$ equal to the year in which the annual report came out. To obtain the calendar year in which a report was filed use the following code:

```
if fyr=0 then delete;
if 1 <= fyr <= 5 then do;
   yyyy = yeara+1;
   end;
else if fyr >=6 then yyyy=yeara;
```

Lets take the example of a firm whose fiscal year end is April. The annual report for fiscal year 1999 is filed sometime around April 30th 2000. Therefore FYR=4, YEARA=1999, and YYYY=2000. Do you count net sales for this filing as sales in calendar year 2000 or calendar year 1999? You have a few choices:

1. The easy way - just use the fiscal year as the "by" variable.

2. Hard way #1 - sum data from calendar quarters. Beware that quarterly data is unaudited and less reliabile than audited annual data.

3. Hard way #2 - assume sales are evenly distributed across months; take (fyr/12) from year t and $((12 - fyr)/12)$ from year $t + 1$. E.g.,

$$\text{sales}_{CY2000} = \frac{4}{12}\text{sales}_{FY1999} + \frac{8}{12}\text{sales}_{FY2000}$$

12.1.2   Quarterly data

With quarterly data, the SPCCQTR and SPCCYR variables can be used. These variables indicate the end of the calendar quarter and year for which data was presented. To convert between fiscal and calendar dates, use a two step process:

1. add YYYYQ to the dataset (YYYYQ = 10*SPCCYR + SPCCQTR)

2. run YYYYQ2SAS macro of Section 3.3.2

12.2   Adjusting for stock splits and stock dividends

Adjust calendar year end stock prices with:

$$P_{adj,t} = \frac{data24}{data27}$$

Adjust calendar year end common shares outstanding with:

$$n_{adj,t} = data25 \times data27$$

12.3   Common financial statement calculations

12.3.1   Book value - common (aka, book value of common equity)

Item	Abbreviation	Annual	Quarterly
total stocholder's equity	*tse*	data216	data60
balance sheet deferred taxes and investment tax credit	*bsdt*	data35	data52
preferred stock carrying value	*pscv*	data130	data55
total assets	*ta*	data6	data44
total liabilities	*tl*	data181	data54
convertible debt	*cd*	n/a	data79

- Method 1: Fama and French [1996]:

"BE is the COMPUSTAT book value of stockholders' equity, plus balance sheet deferred taxes and investment tax credit (if available), minus the book value of preferred stock. Depending on availability, we use redemption, liquidation, or par value (in that order) to estimate the book value of preferred stock. The BE/ME ratio used to form portfolios in June of year t is then book common equity for the fiscal year ending in calendar year $t - 1$, divided by market equity at the end of December of $t - 1$. We do not use negative BE firms, which are rare prior to 1980, when calculating the breakpoints for BE/ME or when forming the size-BE/ME portfolios. Also, only firms with ordinary common equity (as classified by CRSP) are included in the tests. This means that ADR's, REIT's, and units of beneficial interest are excluded."

I interpret this as:

$$BE = tse + bsdt - pscv$$

or, in COMPUSTAT terms:

$$BE = data216 + data35 - data130 \quad \text{(annual)}$$
$$BE = data60 + data52 - data55 \quad \text{(quarterly)}$$

note: Carrying value of preferred stock is used in both equations since quarterly liquidating and redemption preferred stock data are unavailable in COMPUSTAT.

- Method 2: Kayhan and Titman (2003) define book equity as:

$$BE = ta - (tl + pscv) + bsdt + cd$$

However, total assets less total liabilities is the same as total stockholder's equity. Therefore:

$$BE = tse - pscv + bsdt + cd$$

or, in COMPUSTAT terms

$$BE = data216 - data130 + data35 + data79 \quad \text{(annual)}$$
$$BE = data60 - data55 + data52 \quad \text{(quarterly)}$$

Note: convertible debt is unavailable in quarterly data as well as liquidating or redemption value of preferred stock.

The SAS code on WRDS is as follows:

```
data comp1;
   length cusip $8.;
   set comp.compann;
   where data6>0 and data181>0 and not missing (data10)
         and not missing (data35) and not missing (data79);
   BE = sum(Data6,-Data181,-Data10,Data35,Data79);
   cusip = cnum || substr(cic,1,2);
   * Converts fiscal year into calendar year data;
   year = yeara;
   if fyr <= 5 then year= yeara + 1;
   * Accounting data since calendar year 't-1';
   if year >= year(&begindate) - 1 ;
   keep gvkey year fyr BE cusip;
run;
```

12.3.2   Book(total)-to-market value

Chordia and Shivakumar [2006]

> "BM: the natural logarithm of the ratio of the book value of equity plus deferred taxes
> to the market value of equity, using the end of the previous year's market and book
> values. As in Fama and French (1992), the value of BM for July of year t to June of
> year $t + 1$ is computed using accounting data at the end of year $t - 1$, and book-to-
> market ratio values greater than the 0.995 fractile or less than the 0.005 fractile are
> set equal to the 0.995 and 0.005 fractile values, respectively."

Note, in this case, as in Fama and French [1996], portfolios are formed in June of year t. Nevertheless, I interpret this as

$$BM_t = \ln \left[\frac{data216_{t-1} + data35_{t-1}}{MV_t} \right]$$

or, looking at the book value exclusively:

$$BV_t = data216_{t-1} + data35_{t-1} \quad \text{(annual)}$$
$$BV_t = data60_{t-1} + data52_{t-1} \quad \text{(quarterly)}$$

12.3.3   Free cash flow

1. Black [2002] defines free cash flow (FCF) as follows:

$$FCF = \text{sales} - \text{opex} + \text{deprec.} - \text{taxes} - \text{capex}$$

where OPEX is operating expeditures, DEPREC is depreciation, and CAPEX is capital expen-
ditures.

2. Vafeas and Joy [1995] define free cash flow as:

$$FCF = \frac{\text{opinc befor deprec} - \text{int exp} - \text{taxes} - \text{pref and com. div.}}{\text{total assets}}$$

3. Copeland et al. [2005] define free cash flow as:

$$FCF = EBIT\,(1 - \tau_c) + \text{depreciation} - \text{investment}$$

12.4 Output measures

Item	Abbreviation	Annual	Quarterly
net sales (revenue)	r	data12	data2
sale of common and preferred stock (issuance)	si	data108	data84
purchase of common and preferred stock (repurchase)	sp	data115	data93
long term debt issuance	$ltdi$	data111	data86
long term debt repurchase	$ltdr$	data114	data92

12.5 Exclusions and SIC codes

12.5.1 Utilities and financials

Kayhan and Tittman (2007) exclude

1. utilities (SIC codes 4900-4999, NAICS codes 22XXXX)

2. financial firms (SIC codes 6000-6999, NAICS codes 52XXXX)

12.5.2 Small firms

1. Kayhan and Titman [2007] define small firms as firms with book value of assets less than $10 million.

2. Moore [2008] uses the following code to exclude small firms:

```
* ----------------;
* remove bottom 20%;
* ----------------;
proc sort data=cstatdata;
   by date;
run;
proc rank data=cstatdata out=cstatdata groups=5;
   by date;
```

```
    var k;
    ranks rank;
run;
data cstatdata;
    set cstatdata;
    if rank >=1;
run;
```

Chapter 13

CRSP on WRDS

13.1 Dates

CRSP monthly stock file dates represent the end of month day.

13.2 Fields of interest and units

Table 13.1: CRSP.MSF fields of interest
Source: CRSP [2006]

Variable	Description
DATE	date of observation
RET	return including dividends
RETX	return excluding dividends
SHROUT	shares outstanding
PRC	closing price or bid/ask average
	Note: negative numbers indicate reported number is bid/ask average
cfacpr	cumulative factor to adjust prices
cfacshr	cumulative factor to adjust shares outstanding

13.3   Share class and exchange class exclusions

Table 13.2: EXCHCD definitions
Source: CRSP [2006].

Code	Definition
-2	Suspended
-1	Halted
0	Not listed on exchange
1	NYSE
2	AMEX
3	NASDAQ
31	When-issued trading on NYSE
32	When issued trading on AMEX
33	When-issued trading on NASDAQ

Table 13.3: SHRCD definitions
Source: CRSP [2006].

First digit

Code	Definition
1	Ordinary common shares
2	Certificates
3	American Depository Recipts (ADRs)
4	Shares of Beneficial Interest (SBIs)
7	Units (depository, depository receipts, beneficial interest, limited partnership interest, etc.)

Second digit

Code	Definition
0	Securities that have not been further defined
1	Securities that need not be further defined
2	Companies incorporated outside the United States
3	American Trust Components
4	Closed-end funds
5	Closed-end fund companies incorporated outside the US
8	Real Estate Investment Trusts (REITs).

13.4   Return calculation

$$RET_t = \left(\frac{P_t + d_t}{P_{t-1}} \right) - 1$$

where d_t represents dividends paid at time t.

Chapter 14

Interest rates on WRDS

To obtain Federal reserve interest rates from WRDS, use the following macro.

```
%macro tbill(outds);
data &outds;
   set frb.rates_monthly;
   * give me somethin!;
   where not missing(tb_wk4) or not missing (tb_m3)
      or not missing(tb_m6) or not missing(d_tb_y1);
   * note returns are anuualized rates;
   rf1 = tb_wk4/100; * 1-month t-bill rate;
   rf3 = tb_m3/100;  * 3-month t-bill rate;
   rf6 = tb_m6/100;  * 6-month t-bill rate;
   * fill in 1 month t-bill rates;
   if missing(rf1) then rf1=rf3;
   * fill in 6 month t-bill rates;
   if missing(rf6) then rf6=rf3;
   * fill in 12 month t-bill rates;
   if not missing (d_tb_y1) then do;
      rf12 = d_tb_y1/100;
      end;
   else if not missing(rf6) then do;
      rf12 = rf6;
      end;
   else rf12 = rf3;
   keep date rf1 rf3 rf6 rf12;
run;
%mend;
```

Note: output is in total return format (e.g., 0.02 for a 2% rate).

Bibliography

John Black. *Oxford Dictionary of Economics*. Oxford University Press, Oxford New York, second edition, 2002.

A. Colin Cameron and Pravin K. Trivedi. *Microeconometrics: Methods and Applications*. Cambridge, 2005.

Tarun Chordia and Lakshmanan Shivakumar. Earnings and price momentum. *Journal of Financial Economics*, 80:627–656, 2006.

John H. Cochrane. *Asset Pricing*. Princeton University Press, 2005.

Thomas E. Copeland, J. Fred Weston, and Kuldeep Shastri. *Financial Theory and Corporate Policy*. Addison Wesley, 2005.

CRSP. *Data Description Guide*. Center for Research in Security Prices, 105 West Adams Street, Suite 1700, Chicago, IL 60603, ca292.200601 edition, 2006.

Jean-Pierre Danthine and John B. Donaldson. Inflation and asset prices in an exchange economy. *Econometrica*, 54(3):585–605, May 1986.

Lora D. Delwiche and Susan J. Slaughter. *The Little SAS Book*. SAS Institute Inc., Cary, NC, USA, third edition, 2003.

Walter Enders. *Applied Econometric Time Series*. Wiley, 2004.

Eugene F. Fama and Kenneth R. French. Common risk factors in the returns on stocks and bonds. *Journal of Financial Economics*, 33:3–56, 1993.

Eugene F. Fama and Kenneth R. French. Multifactor explanations of asset pricing anomalies. *Journal of Finance*, 51(1):55–84, March 1996.

Wayne E. Ferson and Campbell R. Harvey. Seasonality and consumption-based asset pricing. *Journal of Finance*, 47(2):511–552, 1992.

Michael R. Gibbons, Stephen A. Ross, and Jay Shanken. A test of the efficiency of a given portfolio. *Econometrica*, 57(5):1121–1152, September 1989.

Damodar N. Gujarati. *Basic Econometrics*. McGraw-Hill, 2003.

Lars P. Hansen. Large sample properties of generalized method of moments estimators. *Econometrica*, 50(4):1029–1054, July 1982.

A. Kayhan and S. Titman. Firms' histories and their capital structures. *Journal of Financial Economics*, 83(1):1–32, 2007.

Donald J. Meyer and Jack Meyer. Risk preferences in multi-period consumption models, the euqity premium puzzle, and habit formation utility. *Journal of Monetary Economics*, (52):1497–1515, 2005.

David J. Moore. *Conditional Nonlinear Stochastic Discount Factor Models as Alternative Explanations to Stock Price Momentum*. PhD thesis, University of Tennessee, August 2008.

Seppo Pynnonen. Empirical asset pricing lecture notes - part 2. January 2007.

SAS Institute Inc. *SAS OnlineDoc (R) 9.1.3*. SAS Institute Inc., Cary, North Carolina, 2008.

B. Solnik. The relation between stock prices and inflationary expectations: The international evidence. *Journal of Finance*, 38:35–65, 1983.

Nikos Vafeas and O. Maurice Joy. Open market share repurchases and the free cash flow hypothesis g35. *Economics Letters*, 48:405–410, 1995.